Modular Maths

for Edexcel A/AS Level

Statistics 2

Series Editor Alan Smith

Alan Smith

Anthony Eccles

Alan Graham

Nigel Green

Liam Hennessy

Roger Porkess

S2

Hodder & Stoughton

A MEMBER OF THE HODDER HEADLINE GROUP

Acknowledgements

OCR, AQA and Edexcel accept no responsibility whatsoever for the accuracy or method of working in the answers given.

Orders: please contact Bookpoint Ltd, 78 Milton Park, Abingdon, Oxon OX14 4TD.
Telephone: (44) 01235 827720, Fax: (44) 01235 400454.
Lines are open from 9.00–6.00,
Monday to Saturday, with a 24 hour message answering service.
Email address: orders@bookpoint.co.uk

British Library Cataloguing in Publication Data
A catalogue record for this title is available from The British Library

ISBN 0 340 77993 4

First published 2001
Impression number 10 9 8 7 6 5 4 3 2 1
Year 2006 2005 2004 2003 2002 2001

Typeset by Tech-Set Ltd, Gateshead, Tyne & Wear.
Printed in Great Britain for Hodder & Stoughton Educational, a division of Hodder Headline Plc, 338 Euston Road, London NW1 3BH by J. W. Arrowsmiths Ltd, Bristol.

EDEXCEL ADVANCED MATHEMATICS

The Edexcel course is based on units in the four strands of Pure Mathematics, Mechanics, Statistics and Decision Mathematics. The first unit in each of these strands is designated AS; all others are A2.

The units may be aggregated as follows:

3 units	AS Mathematics
6 units	A Level Mathematics
9 units	A Level Mathematics + AS Further Mathematics
12 units	A Level Mathematics + A Level Further Mathematics

Note that other titles (such as Applied Mathematics, Statistics and so on) are available for certain combinations of units. Full details can be found in the Edexcel Specification booklet.

The six units required for an award in A Level Mathematics must comprise Pure Mathematics 1, 2 and 3, plus three units from the remaining (Applications) strands, including at least one other A2 unit. The synoptic requirement means that a specified pair of units must be examined together in the final session at the end of the course. Again, full details are given in the Edexcel Specification booklet.

Examinations are offered by Edexcel twice a year, in January and in June. Certain units with low candidate numbers (such as Mechanics 6) are offered during the summer sitting only.

Eighteen of the twenty units are assessed by examination only. The exceptions are Statistics 3 and 6, which each contain a project worth 25% of the total marks for that unit.

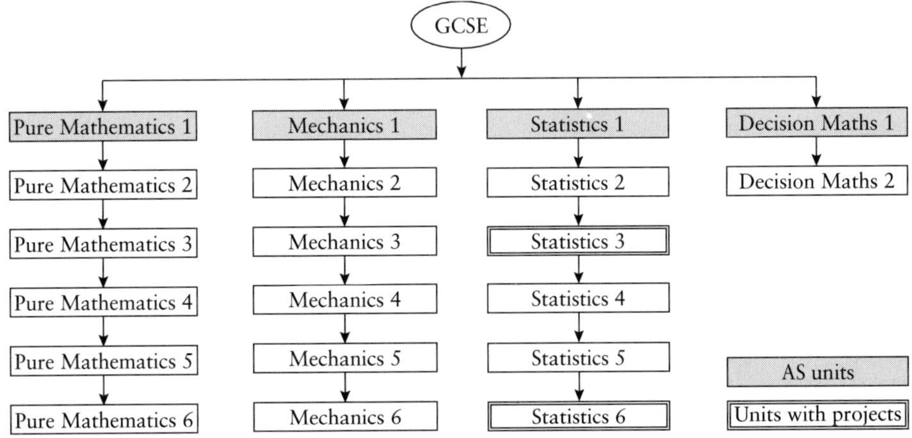

INTRODUCTION

This is the second book in a series written to support the statistics units in the Edexcel Advanced Mathematics scheme. It has been adapted from the successful series written to support the MEI Structured Mathematics schemes and has been substantially edited and rewritten to provide complete coverage of the new Edexcel Pure Mathematics Unit.

There are five chapters in the book. The opening chapter looks at the binomial distribution, solving problems by calculator and also by the use of cumulative tables. This is followed by a similar treatment of the Poisson distribution in Chapter two. Chapters three and four look at continuous probability distributions in some detail, including the use of the normal distribution as an approximating distribution under suitable circumstances. Finally, the book returns to the binomial and Poisson distributions in the context of hypothesis testing.

Each chapter contains a number of past examination and examination-style questions.

Throughout the book you will require the use of a calculator, preferably one with a good selection of statistical functions. You should remember that certain calculator restrictions may be enforced by the examination board, but at the time of writing these do not affect any of the Edexcel statistics units.

I would like to thank the many people who have helped in the presentation and checking of material. Special thanks to Anthony Eccles, Alan Graham, Nigel Green, Liam Hennessy and Roger Porkess who wrote the original editions, and to Terry Heard for his helpful suggestions.

Alan Smith

Contents

Chapter one

THE BINOMIAL DISTRIBUTION

To be or not to be, that is the question.

Shakespeare (Hamlet)

INTRODUCING THE BINOMIAL DISTRIBUTION

Businesswoman Samantha Weeks has invented a new type of cheap energy-saving light bulb. There are still some technical problems at the factory producing these bulbs, and there is a probability of 0.1 that any bulb will be substandard. A substandard bulb will not last as long as it should.

Samantha decides to sell her bulbs in packs of three and believes that if one bulb in a pack is substandard the customers will not complain but that if two or more are substandard they will do so. She also believes that complaints should be kept down to no more than 2.5% of customers. Does she meet her target?

Imagine a pack of Samantha's bulbs. There are eight different ways that good (G) and substandard (S) bulbs can be arranged in Samantha's packs, each with its associated probability.

Arrangement	Probability	Good	Substandard
G G G	$0.9 \times 0.9 \times 0.9 = 0.729$	3	0
G G S	$0.9 \times 0.9 \times 0.1 = 0.081$	2	1
G S G	$0.9 \times 0.1 \times 0.9 = 0.081$	2	1
S G G	$0.1 \times 0.9 \times 0.9 = 0.081$	2	1
G S S	$0.9 \times 0.1 \times 0.1 = 0.009$	1	2
S G S	$0.1 \times 0.9 \times 0.1 = 0.009$	1	2
S S G	$0.1 \times 0.1 \times 0.9 = 0.009$	1	2
S S S	$0.1 \times 0.1 \times 0.1 = 0.001$	0	3

Putting these results together gives this table:

Good	Substandard	Probability
3	0	0.729
2	1	0.243
1	2	0.027
0	3	0.001

So the probability of more than one substandard bulb in a pack is

$$0.027 + 0.001 = 0.028 \text{ or } 2.8\%$$

This is slightly more than the 2.5% that Samantha regards as acceptable.

What business advice would you give Samantha?

In this example we wrote down all the possible outcomes and found their probabilities one at a time, as you do in a tree diagram. Even with just three bulbs this was repetitive. If Samantha had packed her bulbs in boxes of six it would have taken 64 lines to list them all. Clearly we need to find a less cumbersome approach.

You will have noticed that in the case of two good bulbs and one substandard, the probability is the same for each of the three arrangements in the box.

Arrangement	Probability	Good	Substandard
G G S	$0.9 \times 0.9 \times 0.1 = 0.081$	2	1
G S G	$0.9 \times 0.1 \times 0.9 = 0.081$	2	1
S G G	$0.1 \times 0.9 \times 0.9 = 0.081$	2	1

So the probability of this outcome is $3 \times 0.081 = 0.243$. The number 3 arises because there are three ways of arranging two good and one substandard bulb in the box.

EXAMPLE 1.1

How many different ways are there of arranging the letters *GGS*?

Solution

Since all the letters are either *G* or *S*, all you need to do is to count the number of ways of choosing the letter *G* two times out of three letters. This is

$${}^3C_2 = \frac{3!}{2! \times 1!} = \frac{6}{2} = 3$$

So what does this tell you? There was no need to list all the possibilities for Samantha's boxes of bulbs. The information could have been written down like this:

Good	Substandard	Expression	Probability
3	0	${}^3C_3\,(0.9)^3$	0.729
2	1	${}^3C_2\,(0.9)^2(0.1)^1$	0.243
1	2	${}^3C_1\,(0.9)^1(0.1)^2$	0.027
0	3	${}^3C_0\,(0.1)^3$	0.001

Samantha's light bulbs are an example of a common type of situation which is modelled by the binomial distribution. In describing such situations in this book, we emphasise the fact by using the word trial rather than the more general term experiment.

- You are conducting trials on random samples of a certain size, denoted by n.
- There are just two possible outcomes (in this case substandard and good). These are often referred to as *success* and *failure*.
- Both outcomes have fixed probabilities, the two adding to 1. The probability of success is usually called p, that of failure q, so $p + q = 1$.
- The probability of success in any trial is independent of the outcomes of previous trials.

You can then list the probabilities of the different possible outcomes as in the table above.

The method of the previous section can be applied more generally. You can call the probability of a substandard bulb p (instead of 0.1), the probability of a good bulb q (instead of 0.9) and the number of substandard bulbs in a packet of three, X.

Then the possible values of X and their probabilities are as shown in the table below.

X	0	1	2	3
Probability	q^3	$3pq^2$	$3p^2q$	p^3

This package of values of X with their associated probabilities is called a *probability distribution*.

If Samantha decided to put five bulbs in a packet the probability distribution would be

X	0	1	2	3	4	5
Probability	q^5	$5pq^4$	$10p^2q^3$	$10p^3q^2$	$5p^4q$	p^5

10 is ${}^5C_2$

The entry for $X = 2$, for example, arises because there are two 'successes' (substandard bulbs), giving probability p^2, and three 'failures' (good bulbs), giving probability q^3, and these can happen in ${}^5C_2 = 10$ ways. This can be written as $P(X = 2) = 10p^2q^3$.

You should already be familiar with the binomial theorem from *Pure Mathematics 1*. You will notice that the probabilities in the table are the terms of the binomial expansion of $(q + p)^5$. This is why this is called a binomial distribution. Notice also that the sum of these probabilities is $(q + p)^5 = 1^5 = 1$, since $q + p = 1$, which is to be expected since the distribution covers all possible outcomes.

THE GENERAL CASE

The general binomial distribution deals with the possible numbers of successes when there are n trials, each of which may be a success (with probability p) or a failure (with probability q); p and q are fixed positive numbers and $p + q = 1$. This distribution is denoted by $B(n, p)$. So, the original probability distribution for the number of substandard bulbs in Samantha's boxes of three is $B(3, 0.1)$.

For $B(n, p)$, the probability of r successes in n trials is found by the same argument as before. Each success has probability p and each failure has probability q, so the probability of r successes and $(n - r)$ failures in a particular order is $p^r q^{n-r}$. The positions in the sequence of n trials which the successes occupy can be chosen in ${}^nC_r$ ways. Therefore

$$P(X = r) = {}^nC_r p^r q^{n-r} \quad \text{for} \quad 0 \leqslant r \leqslant n$$

The successive probabilities for $X = 0, 1, 2, \ldots, n$ are the terms of the binomial expansion of $(q + p)^n$.

Notes

1 The number of successes, X, is a variable which takes a restricted set of values ($X = 0, 1, 2, \ldots, n$) each of which has a known probability of occurring. This is an example of a *random variable*. Random variables are usually denoted by upper case letters, such as X, but the particular values they may take are written in lower case, such as r. To state that X has the binomial distribution $B(n, p)$ you can use the abbreviation $X \sim B(n, p)$, where the symbol $\sim$ means 'has the distribution'.

2 It is often the case that you use a theoretical distribution, such as the binomial, to describe a random variable that occurs in real life. This process is called modelling and it enables you to carry out relevant calculations. If the theoretical distribution matches the real life variable perfectly, then the model is perfect. Usually, however, the match is quite good but not perfect. In this case the results of any calculations will not necessarily give a completely accurate description of the real life situation. They may, nonetheless, be very useful.

Exercise 1A

1 The recovery ward in a maternity hospital has six beds. What is the probability that the mothers there have between them four girls and two boys? (You may assume that there are no twins and that a baby is equally likely to be a girl or a boy.)

2 A typist has a probability of 0.99 of typing a letter correctly. He makes his mistakes at random. He types a sentence containing 200 letters. What is the probability that he makes exactly one mistake?

3 In a well-known game you have to decide which your opponent is going to choose: 'Paper', 'Stone' or 'Scissors'. If you guess entirely at random, what is the probability that you are right exactly 5 times out of 15?

4 There is a fault in a machine making microchips, with the result that only 80% of those it produces work. A random sample of eight microchips made by this machine is taken. What is the probability that exactly six of them work?

5 An airport is situated in a place where poor visibility (less than 800 m) can be expected 25% of the time. A pilot flies into the airport on ten different occasions. What is the probability that he encounters poor visibility exactly four times?

6 Three coins are tossed.
(a) What is the probability of all three showing heads?
(b) What is the probability of two heads and one tail?
(c) What is the probability of one head and two tails?
(d) What is the probability of all three showing tails?
(e) Show that the probabilities for the four possible outcomes add up to 1.

7 A coin is tossed ten times.
(a) What is the probability of it coming down heads five times and tails five times?
(b) Which is more likely: exactly seven heads or more than seven heads?

8 In an election 30% of people support the Progressive Party. A random sample of eight voters is taken.
(a) What is the probability that it contains
(i) 0 **(ii)** 1 **(iii)** 2 **(iv)** at least 3 supporters of the Progressive Party?
(b) Which is the most likely number of Progressive Party supporters to find in a sample of size eight?

9 There are 15 children in a class.
(a) What is the probability that
(i) 0 **(ii)** 1 **(iii)** 2 **(iv)** at least 3 were born in January?
(b) What assumption have you made in answering this question? How valid is this assumption in your view?

10 Criticise this argument.

If you toss two coins they can come down three ways: two heads, one head and one tail, or two tails. There are three outcomes and so each of them must have probability one third.

THE MEAN AND VARIANCE OF B(n, p)

It can be shown algebraically that the mean (or expectation) of the binomial distribution B(n, p) is np and the variance is $np(1 - p)$ or npq.

You do not need to know the theory of the algebraic derivation of these results, but you are expected to know how to use them in solving problems.

EXAMPLE 1.2

The number of substandard bulbs, X, in a packet of three bulbs produced at Samantha's factory is modelled by the binomial distribution B(3, 0.1).

(a) Find the mean number of substandard bulbs per packet.

(b) Find the variance of the number of substandard bulbs per packet.

Solution

(a) Mean $= np = 3 \times 0.1 = 0.3$

(b) Variance $= npq = 3 \times 0.1 \times 0.9 = 0.27$

Note

Although X can only take the integer values 0, 1, 2, 3, its mean is a decimal, 0.3. It would be wrong to try to round off the mean to an integer value.

EXERCISE 1B

1 An ordinary six-sided die is thrown 36 times. The random variable X denotes the number of sixes which are recorded out of the 36 throws.

(a) Write down the distribution of X.

(b) Find the mean of X.

(c) Find the variance of X, and hence obtain its standard deviation.

2 The random variable $X \sim$ B(16, 0.25).

(a) Find $P(X = 5)$.

(b) Find the mean of X.

(c) Find the variance of X.

3 In a multiple-choice exam each question has five possible answers, marked A, B, C, D, E; only one of these five answers is correct. A nervous candidate guesses the answers to eight questions at random.

(a) Find the mean number of questions he guesses correctly.

(b) Find the variance.

4 $X \sim B(6, 0.9)$.
  (a) Find $P(X = 4)$.
  (b) Find the mean of X.
  (c) Find the variance of X.

5 An amateur rocket scientist reckons that 8% of his launches are failures. He plans to launch his rocket 40 times during the coming summer.
  (a) Explain why a binomial distribution might be used to model the number of failures, X, out of the 40 launches.
  (b) Find the mean of X.
  (c) Find the variance of X.
  (d) Find the probability that exactly 35 launches are successful.

6 The random variable X is known to be binomially distributed with $n = 12$. The mean of X is 9. Find p.

7 The random variable X is known to be binomially distributed with $p = 0.3$. The expectation of X is 15. Find n.

8 The random variable X is known to be binomially distributed with $n = 20$. The variance of X is 3.2. Find the value of p.

USE OF CUMULATIVE BINOMIAL TABLES

Some problems require you to calculate several binomial probabilities and add them together. Such problems can often be shortened by using cumulative binomial probabilities, which are tabulated in the Edexcel book of statistical tables and formulae.

EXAMPLE 1.3

A pack of cards contains 13 each of the suits Clubs, Hearts, Diamonds, Spades. A card is selected at random, then replaced. This is repeated a total of eight times. Find the probability of obtaining at most two Spades.

Solution ***Method 1***

Let X denote the number of Spades obtained from eight trials. $X \sim B(8, 0.25)$

$$P(X = 0) = {}^8C_0 \times 0.25^0 \times 0.75^8 = 0.1001$$

$$P(X = 1) = {}^8C_1 \times 0.25^1 \times 0.75^7 = 0.2670$$

$$P(X = 2) = {}^8C_2 \times 0.25^2 \times 0.75^6 = 0.3115$$

$$\text{Thus } P(X \leqslant 2) = 0.1001 + 0.2670 + 0.3115 = 0.6786$$

Method 2

Let X denote the number of Spades obtained from eight trials.

$X \sim B(8, 0.25)$

Consult cumulative binomial tables.

Binomial cumulative distribution function

The tabulated value is $P(X \leqslant x)$, where X has a binomial distribution with index n and parameter p.

$p =$	0.05	0.10	0.15	0.20	0.25	0.30	0.35	0.40	0.45	0.50
$n = 5, x = 0$	0.7738	0.5905	0.4437	0.3277	0.2373	0.1681	0.1160	0.0778	0.0503	0.0312
1	0.9774	0.9185	0.8352	0.7373	0.6328	0.5282	0.4284	0.3370	0.2562	0.1875
2	0.9988	0.9914	0.9734	0.9421	0.8965	0.8369	0.7648	0.6826	0.5931	0.5000
3	1.0000	0.9995	0.9978	0.9933	0.9844	0.9692	0.9460	0.9130	0.8688	0.8125
4	1.0000	1.0000	0.9999	0.9997	0.9990	0.9976	0.9947	0.9898	0.9815	0.9688
$n = 6, x = 0$	0.7351	0.5314	0.3771	0.2621	0.1780	0.1176	0.0754	0.0467	0.0277	0.0156
1	0.9672	0.8857	0.7765	0.6554	0.5339	0.4202	0.3191	0.2333	0.1636	0.1094
2	0.9978	0.9842	0.9527	0.9011	0.8306	0.7443	0.6471	0.5443	0.4415	0.3438
3	0.9999	0.9987	0.9941	0.9830	0.9624	0.9295	0.8826	0.8208	0.7447	0.6563
4	1.0000	0.9999	0.9996	0.9984	0.9954	0.9891	0.9777	0.9590	0.9308	0.8906
5	1.0000	1.0000	1.0000	0.9999	0.9998	0.9993	0.9982	0.9959	0.9917	0.9844
$n = 7, x = 0$	0.6983	0.4783	0.3206	0.2097	0.1335	0.0824	0.0490	0.0280	0.0152	0.0078
1	0.9556	0.8503	0.7166	0.5767	0.4449	0.3294	0.2338	0.1586	0.1024	0.0625
2	0.9962	0.9743	0.9262	0.8520	0.7564	0.6471	0.5323	0.4199	0.3164	0.2266
3	0.9998	0.9973	0.9879	0.9667	0.9294	0.8740	0.8002	0.7102	0.6083	0.5000
4	1.0000	0.9998	0.9988	0.9953	0.9871	0.9712	0.9444	0.9037	0.8471	0.7734
5	1.0000	1.0000	0.9999	0.9996	0.9987	0.9962	0.9910	0.9812	0.9643	0.9375
6	1.0000	1.0000	1.0000	1.0000	0.9999	0.9998	0.9994	0.9984	0.9963	0.9922
$n = 8, x = 0$	0.6634	0.4305	0.2725	0.1678	0.1001	0.0576	0.0319	0.0168	0.0084	0.0039
1	0.9428	0.8131	0.6572	0.5033	0.3671	0.2553	0.1691	0.1064	0.0632	0.0352
2	0.9942	0.9619	0.8948	0.7969	0.6785	0.5518	0.4278	0.3154	0.2201	0.1445
3	0.9996	0.9950	0.9786	0.9437	0.8862	0.8059	0.7064	0.5941	0.4770	0.3633
4	1.0000	0.9996	0.9971	0.9896	0.9727	0.9420	0.8939	0.8263	0.7396	0.6367
5	1.0000	1.0000	0.9998	0.9988	0.9958	0.9887	0.9747	0.9502	0.9115	0.8555
6	1.0000	1.0000	1.0000	0.9999	0.9996	0.9987	0.9964	0.9915	0.9819	0.9648
7	1.0000	1.0000	[illegible]	[illegible]	1.0000	[illegible]	[illegible]	0.9993	[illegible]	0.9961

Figure 1.1

Thus $P(X \leqslant 2) = 0.6785$

Note

The slight discrepancy between these two methods arises from rounding errors in the three separate calculations within Method 1.

Although the cumulative binomial tables are very useful, they cannot list every possible binomial distribution, so you do need to be prepared to work out all the separate cases by longhand if necessary.

Notice also that the tabulated values of p stop at 0.5. For values above this you need to work with $1-p$, as shown in the next example.

EXAMPLE 1.4

Michael is a talented football player. His speciality is taking penalty spot kicks, and he reckons that he can score 85% of the time, on average. In a typical season he takes 25 penalties.

Using the model B(25, 0.85) find the probability that

(a) he scores at least 20 times

(b) he scores less than 18 times.

Solution

Let X be the number of times Michael scores out of 25 attempts. Then $X \sim \mathrm{B}(25, 0.85)$.

Let Y be the number of times he does *not* score out of 25 attempts. Then $Y \sim \mathrm{B}(25, 0.15)$.

(a) $\mathrm{P}(X \text{ is at least } 20) = \mathrm{P}(Y \leqslant 5)$
$= 0.8385$ (from tables)

(b) $\mathrm{P}(X < 18) = \mathrm{P}(Y > 7)$
$= 1-\mathrm{P}(Y \leqslant 6)$
$= 1-0.9305$
$= 0.0695$

Notice how this works at the upper tail of the distribution: to find the probability that $Y > 7$ you must compute the probability of $Y \leqslant 6$ from tables, and then subtract it from 1.

	$p =$	0.05	0.10	0.15	0.20	0.25	0.30	0.35	0.40	0.45	0.50
$n = 25$,	$x = 0$	0.2774	0.0718	0.0172	0.0038	0.0008	0.0001	0.0000	0.0000	0.0000	0.0000
	1	0.6424	0.2712	0.0931	0.0274	0.0070	0.0016	0.0003	0.0001	0.0000	0.0000
	2	0.8729	0.5371	0.2537	0.0982	0.0321	0.0090	0.0021	0.0004	0.0001	0.0000
	3	0.9659	0.7636	0.4711	0.2340	0.0962	0.0332	0.0097	0.0024	0.0005	0.0001
	4	0.9928	0.9020	0.6821	0.4207	0.2137	0.0905	0.0320	0.0095	0.0023	0.0005
	5	0.9988	0.9666	0.8385	0.6167	0.3783	0.1935	0.0826	0.0294	0.0086	0.0020
	6	0.9998	0.9905	0.9305	0.7800	0.5611	0.3407	0.1734	0.0736	0.0258	0.0073
	7	1.0000	0.9977	0.9745	0.8909	0.7265	0.5118	0.3061	0.1536	0.0639	0.0216
	8	1.0000	0.9995	0.9920	0.9532	0.8506	0.6769	0.4668	0.2735	0.1340	0.0539
	9	1.0000	0.9999	0.9979	0.9827	0.9287	0.8106	0.6303	0.4246	0.2424	0.1148
	10	1.0000	1.0000	[illegible]		[illegible]	[illegible]		[illegible]	[illegible]	

FIGURE 1.2

EXAMPLE 1.5

Extensive research has shown that 1 person out of every 4 is allergic to a particular grass seed. A group of 20 university students volunteer to try out a new treatment.

(a) What is the mean of the number of allergic people in the group?

(b) What is the probability that **(i)** exactly two **(ii)** no more than two of the group are allergic?

(c) How large a sample would be needed for the probability of it containing at least one allergic person to be greater than 99.9%.

(d) What assumptions have you made in your answer?

Solution

This situation is modelled by the binomial distribution with $n = 20$, $p = 0.25$ and $q = 0.75$. The number of allergic people is denoted by X.

(a) Mean $= np = 20 \times 0.25 = 5$ people.

(b) $X \sim \mathrm{B}(20, 0.25)$

(i) $\mathrm{P}(X = 2) = {}^{20}\mathrm{C}_2(0.75)^{18}(0.25)^2 = 0.067$

(ii) $\mathrm{P}(X \leqslant 2) = 0.0913$ (from tables)

(c) Let the sample size be n (people), so that $X \sim \mathrm{B}(n, 0.25)$.

The probability that none of them is allergic is

$$\mathrm{P}(X = 0) = (0.75)^n$$

and so the probability that at least one is allergic is

$$\mathrm{P}(X \geqslant 1) = 1 - \mathrm{P}(X = 0) = 1 - (0.75)^n$$

So we need $1 - (0.75)^n > 0.999$

$$(0.75)^n < 0.001$$

Solving $(0.75)^n = 0.001$

gives $n \log 0.75 = \log 0.001$

$$n = \log 0.001 \div \log 0.75$$

$$= 24.01$$

So 25 people are required.

Note

Although 24.01 is very close to 24 it would be incorrect to round down.
$1 - (0.75)^{24} = 0.998\,996\,6$ which is just less than 99.9%.

(d) The assumptions made are:

(i) That the sample is random. This is almost certainly untrue. University students are nearly all in the 18–25 age range and so a sample of them cannot be a random sample of the whole population. They may well also be unrepresentative of the whole population in other ways. Volunteers are seldom truly random.

(ii) That the outcome for one person is independent of that for another. This is probably true unless they are a group of friends from, say, an athletics team, where those with allergies are less likely to be members.

EXERCISE 1C

1 During a promotional offer, cereal packets contain a toy cat, dog, fish or horse. Each of the four animals is equally likely.

(a) Anna buys six boxes of cereal at random. Find the probability that she gets exactly two fish.

(b) Bradley buys ten boxes of cereal at random. Find the probability that he gets no more than four fish.

2 It is known that $Y \sim B(9, 0.4)$.

(a) Calculate $P(Y = 7)$.

(b) Use tables to find $(Y > 6)$.

3 In a large bird colony it is known that 55% of the adult birds are female. Twenty birds are trapped at random.

(a) Find the probability that exactly 14 are female.

(b) Find the mean and the variance of the number of females trapped.

(c) Find the probability that the number of females trapped is less than the mean.

4 The random variable $X \sim B(22, 0.12)$. Find the probability that

(a) $X = 0$

(b) $X = 2$

(c) $X \leqslant 2$.

5 A railway company reckons that 88% of its trains run on time. An inspector decides he is going to check the timekeeping of a random sample of ten trains. Assuming the figure of 88% to be correct:

(a) calculate the probability that exactly eight of the ten trains run on time.

In fact, it is subsequently discovered that the correct figure is 90%. Using this new value, find the probability that:

(b) exactly eight of the ten trains run on time

(c) at least eight of the ten trains run on time.

6 Five coins are thrown together. The random variable X represents the number of heads showing when the coins land.

(a) Explain briefly why a binomial model might be appropriate for X. State the values of any parameters required.

(b) Calculate $P(X = 3)$.

(c) Use tables to find $P(X \leqslant 3)$.

7 The random variable $X \sim B(20, 0.75)$. Find the probability that

(a) $X = 14$

(b) X is less than 14

(c) X is at least 17.

8 In a particular area 30% of men and 20% of women are overweight and there are four men and three women working in an office there. Find the probability that there are

(a) 0 **(b)** 1 **(c)** 2 overweight men;

(d) 0 **(e)** 1 **(f)** 2 overweight women;

(g) 2 overweight people in the office.

What assumption have you made in answering this question?

9 Bella Cicciona, a fortune teller, claims to be able to predict the sex of unborn children. In fact, on each occasion she is consulted, the probability that she makes a correct prediction is 0.6, independent of any other prediction.

One afternoon, Bella is consulted by ten expectant mothers. Find, correct to 2 significant figures, the probabilities that

(a) her first eight predictions are correct and her last two are wrong

(b) she makes exactly eight correct predictions

(c) she makes at least eight correct predictions

(d) she makes exactly eight correct predictions given that she makes at least eight.

[MEI]

10 A company uses machines to manufacture wine glasses. Because of imperfections in the glass it is normal for 10% of the glasses to leave the machine cracked. The company takes regular samples of size 10 from each machine. If more than 2 glasses in a sample are cracked, they stop the machine and check that it is set correctly.

(a) What is the probability that a sample of size 10 contains

(i) 0 **(ii)** 1 **(iii)** 2 faulty glasses, when the machine is correctly set?

(b) What is the probability that as a result of taking a sample a machine is stopped when it is correctly set?

(c) A machine is in fact incorrectly set and 20% of the glasses from it are cracked. What is the probability that this is undetected by a particular sample?

EXERCISE 1D Examination-style questions

1 In a game five dice are rolled together.

(a) What is the probability that

(i) all five show 1 **(ii)** exactly three show 1 **(iii)** none of them shows 1?

(b) What is the most likely number of times for 6 to show?

2 There are eight colours of Smarties which normally occur in equal proportions: red, orange, yellow, green, blue, purple, pink and brown. Veronica's mother gives each of her children 16 Smarties. Veronica says that the blue ones are much nicer than the rest and is very upset when she receives less than her fair share of them.

(a) How many blue Smarties did Veronica expect to get?

(b) What was the probability that she would receive fewer blue ones than she expected?

(c) What was the probability that she would receive more blue ones than she expected?

3 On her drive to work Stella has to go through four sets of traffic lights. She estimates that for each set the probability of her finding them red is $\frac{2}{3}$ and green $\frac{1}{3}$. (She ignores the possibility of them being amber.) Stella also estimates that when a set of lights is red she is delayed by one minute.

(a) Find the probability of

(i) 0 **(ii)** 1 **(iii)** 2 **(iv)** 3 sets of lights being against her.

(b) Find the expected extra journey time due to waiting at lights.

4 A drunken man steps out of a bar into a narrow alley which runs from west to east. At each step he chooses at random whether to go east or west. After 12 steps he stops for a rest.

(a) Why is it impossible for him then to be 1 step away from the bar?

(b) What is the probability that he is then 10 steps east of the bar?

(c) What is his most likely position?

(d) What is the probability that he is 4 steps away from the bar, either to the east or to the west?

5 Pepper moths are found in two varieties, light and dark. The proportion of dark moths increases with certain types of atmospheric pollution. At the time of the question 30% of the moths in a particular town are dark.
A research student sets a moth trap and catches nine moths, four light and five dark.

(a) What is the probability of that result for a sample of nine moths?

(b) What is the expected number of dark moths in a sample of nine?

The next night the student's trap catches ten pepper moths.

(c) What is the probability that the number of dark moths in this sample is the same as the expected number?

6 An insurance salesman sells policies to five men, all of identical age and in good health. According to his company's records the probability that a man of this particular age will be alive in 20 years' time is 0.4. Find the probability that in 20 years' time the number of men still alive will be

(a) five
(b) at least three
(c) exactly two
(d) at least one.

[Cambridge, adapted]

7 A general knowledge quiz has ten questions. Each question has three possible 'answers' of which one only is correct. A woman attempts the quiz by pure guesswork.

(a) Find the probabilities that she obtains

(i) exactly two correct answers **(ii)** not more than two correct answers.

(b) What is the most likely number of correct answers and the probability that she just achieves this number?

[MEI]

8 Six fair coins are tossed and those landing heads uppermost are eliminated. The remainder are tossed again and the process of elimination is repeated. Tossing and elimination continue in this way until no coins are left.

Find the probabilities of the following events.

(a) All six coins are eliminated in the first round.
(b) Exactly two coins are eliminated in the first round.
(c) Exactly two coins are eliminated in the first round and exactly two coins are eliminated in the second round.
(d) Exactly two coins are eliminated in *each* of the first three rounds.
(e) Exactly two coins are eliminated in the first round and exactly two coins are eliminated in the *third* round.

[MEI]

9 A multiple choice test consists of 20 questions. There are five choices for each question and only one choice is correct. A certain candidate made a random guess for each question.

(a) Suggest a suitable model to describe the number of questions the candidate guessed correctly and give suitable values for any parameters required.

Find the probability that the candidate obtained

(b) no correct answers
(c) more than seven correct answers.

[Edexcel]

10 A simple fairground game consists of a rectangular target divided into 20 equally-sized squares. Seven of these squares are coloured red and the rest are of other colours. Players are blindfolded and given ten darts to throw at the board. Prizes are given for darts which land in red squares. Any dart which misses the board is returned and is repeatedly thrown until it hits the board.

(a) Suggest a suitable model to describe the number of darts that land in red squares and give the values of any parameters required.

Find the probability that a player obtains

(b) fewer than three darts in red squares

(c) at least six darts in red squares.

A skilled darts player asks if he can play the game without the blindfold.

(d) Explain briefly what feature of the model in part **(a)** will need refinement.

[Edexcel]

KEY POINTS

The binomial distribution may be used to model situations in which:

1. you are conducting trials on random samples of a certain size, n
2. there are two possible outcomes, often referred to as success and failure
3. both outcomes have fixed probabilities, p and q, and $p + q = 1$
4. the probability of success in any trial is independent of the outcomes of previous trials
5. The probability that the number of successes, X, has the value r, is given by

$$P(X = r) = {}^nC_r p^r q^{n-r}.$$

6. For $B(n, p)$ the expectation of the number of successes is np.

To be and not to be, that is the answer.

Piet Hein

Chapter two

The Poisson distribution

If something can go wrong, sooner or later it will go wrong.

Murphy's Law

Introducing the Poisson distribution

The Poisson distribution was developed early in the nineteenth century. It is a powerful distribution in its own right, and may also be used, under certain circumstances, as an approximation to the binomial distribution.

The Poisson distribution is generated as follows:

$$P(X = r) = \frac{e^{-\lambda}\lambda^r}{r!} \qquad r = 0, 1, 2, 3, \ldots$$

where λ (lambda) is a positive number known as the *parameter* for that particular distribution.

To see where this formula comes from, we shall consider a problem concerning the binomial distribution, and develop an approximate solution to it.

Suppose a rare disease is known to affect, on average, 1 person in every 40 000. In a town of 60 000 residents, how likely is it that, say, 5 people would suffer from the disease?

The situation could be modelled by the binomial distribution. The probability of somebody getting the disease in any year is $\frac{1}{40\,000}$ and so that of not getting it is $1 - \frac{1}{40\,000} = \frac{39\,999}{40\,000}$.

The probability of 5 cases among 60 000 people (and so 59 995 people not getting the disease) is given by

$$^{60\,000}C_5 \left(\frac{39\,999}{40\,000}\right)^{59\,995} \left(\frac{1}{40\,000}\right)^5 \approx 0.0141$$

What you really want to know, however, is not the probability of exactly 5 cases but that of 5 or more cases.

You can find the probability of 5 or more cases by finding the probability of up to and including 4 cases, and subtracting it from 1.

The probability of up to and including 4 cases is given by:

$$\left(\frac{39\,999}{40\,000}\right)^{60\,000} \qquad \text{0 cases}$$

$$+ \, {}^{60\,000}C_1 \left(\frac{39\,999}{40\,000}\right)^{59\,999} \left(\frac{1}{40\,000}\right) \qquad \text{1 case}$$

$$+ \, {}^{60\,000}C_2 \left(\frac{39\,999}{40\,000}\right)^{59\,998} \left(\frac{1}{40\,000}\right)^2 \qquad \text{2 cases}$$

$$+ \, {}^{60\,000}C_3 \left(\frac{39\,999}{40\,000}\right)^{59\,997} \left(\frac{1}{40\,000}\right)^3 \qquad \text{3 cases}$$

$$+ \, {}^{60\,000}C_4 \left(\frac{39\,999}{40\,000}\right)^{59\,996} \left(\frac{1}{40\,000}\right)^4 \qquad \text{4 cases}$$

It is messy but you can evaluate it on your calculator. It comes out to be

$$0.223 + 0.335 + 0.251 + 0.126 + 0.047 = 0.981$$

(The figures are written to three decimal places but more places were used in the calculation.)

So the probability of 5 or more cases in a year is $1 - 0.981 = 0.019$. It is unlikely but certainly could happen, see figure 2.1.

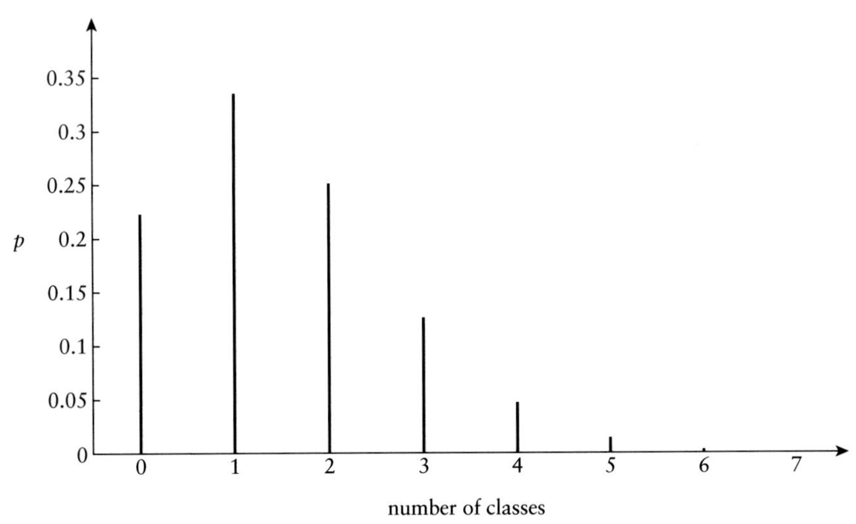

FIGURE 2.1 *Probability distribution* $B(60\,000, \frac{1}{40\,000})$

Note The binomial model assumes the trials are independent. If this disease is at all infectious, that certainly would not be the case.

Approximating the binomial terms

Although it was possible to do the calculation using results derived from the binomial distribution, it was distinctly cumbersome. In this section you will see how the calculations can be simplified, a process which turns out to be unexpectedly profitable. The work that follows depends upon the facts that the event is rare but there are many opportunities for it to occur: that is, p is small and n is large.

Start by looking at the first term, the probability of 0 cases of the disease. This is

$$\left(\frac{39\,999}{40\,000}\right)^{60\,000} = k, \text{ a constant}$$

Now look at the next term, the probability of 1 case of the disease. This is

$$^{60\,000}C_1\left(\frac{39\,999}{40\,000}\right)^{59\,999}\left(\frac{1}{40\,000}\right)$$

$$= \frac{60\,000 \times \left(\frac{39\,999}{40\,000}\right)^{60\,000} \times \left(\frac{40\,000}{39\,999}\right)}{40\,000}$$

$$= k \times \frac{60\,000}{39\,999} \approx k \times \frac{60\,000}{40\,000} = k \times 1.5$$

Now look at the next term, the probability of 2 cases of the disease. This is

$$^{60\,000}C_2 \times \left(\frac{39\,999}{40\,000}\right)^{59\,998} \times \left(\frac{1}{40\,000}\right)^2$$

$$= \frac{60\,000 \times 59\,999}{2 \times 1} \times \left(\frac{39\,999}{40\,000}\right)^{60\,000} \times \left(\frac{40\,000}{39\,999}\right)^2 \times \left(\frac{1}{40\,000}\right)^2$$

$$= \frac{k \times 60\,000 \times 59\,999}{2 \times 1 \times 39\,999 \times 39\,999} \approx \frac{k \times 60\,000 \times 60\,000}{2 \times 40\,000 \times 40\,000} = k \times \frac{(1.5)^2}{2}$$

Proceeding in this way leads to the following probability distribution for the number of cases of the disease:

Number of cases	0	1	2	3	4	...
Probability	k	$k \times 1.5$	$\frac{k \times (1.5)^2}{2!}$	$\frac{k \times (1.5)^3}{3!}$	$\frac{k \times (1.5)^4}{4!}$	...

Since the sum of the probabilities = 1,

$$k + k \times 1.5 + k \times \frac{(1.5)^2}{2!} + k \times \frac{(1.5)^3}{3!} + k \times \frac{(1.5)^4}{4!} + \cdots = 1$$

$$k\left[1 + 1.5 + \frac{(1.5)^2}{2!} + \frac{(1.5)^3}{3!} + \frac{(1.5)^4}{4!} + \cdots\right] = 1$$

The terms in the square brackets form a well-known series in pure mathematics, the exponential series e^x.

$$e^x = 1 + x + \frac{x^2}{2!} + \frac{x^3}{3!} + \frac{x^4}{4!} + \cdots$$

Since $k \times e^{1.5} = 1$, $k = e^{-1.5}$.

This gives the probability distribution for the number of cases of the disease:

Number of cases	0	1	2	3	4	...
Probability	$e^{-1.5}$	$e^{-1.5}1.5$	$e^{-1.5}\frac{(1.5)^2}{2!}$	$e^{-1.5}\frac{(1.5)^3}{3!}$	$e^{-1.5}\frac{(1.5)^4}{4!}$	...

and in general for r cases the probability is $e^{-1.5}\frac{(1.5)^r}{r!}$.

ACCURACY

These expressions are clearly much simpler than those involving binomial coefficients. How accurate are they? The following table compares the results from the two methods, given to six decimal places.

	Probability	
No. of cases	**Exact binomial method**	**Approximate method**
0	0.223 126	0.223 130
1	0.334 697	0.334 695
2	0.251 025	0.251 021
3	0.125 512	0.125 511
4	0.047 066	0.047 067

You will see that the agreement is very good; there are no differences until the sixth decimal places.

THE POISSON DISTRIBUTION

What started out as a search for an easy way to calculate terms in a binomial distribution has ended up with terms which are so different that they are seen as a completely different distribution, called the *Poisson distribution*.

In the example the distribution had mean 1.5 and the general term was given by $P(X = r) = e^{-1.5}\frac{(1.5)^r}{r!}$ where the discrete random variable X denotes the number of cases of the disease.

This can be generalised to the Poisson distribution with mean λ for which

$$P(X = r) = e^{-\lambda}\frac{\lambda^r}{r!}.$$

Notes

1. The mean, λ, is usually called the population parameter.
2. e is the base of natural logarithms. e = 2.718 28. . .; like π, it does not terminate. e^x is a function on your calculator. It is sometimes written $\exp(x)$.
3. The upper case letter X is a random variable, and the lower case r is a particular value that X can take.
4. The shape of the Poisson distribution depends on the value of the parameter, λ. If λ is small the distribution has positive skew, but as λ increases the distribution becomes progressively more symmetrical, see figure 2.2.

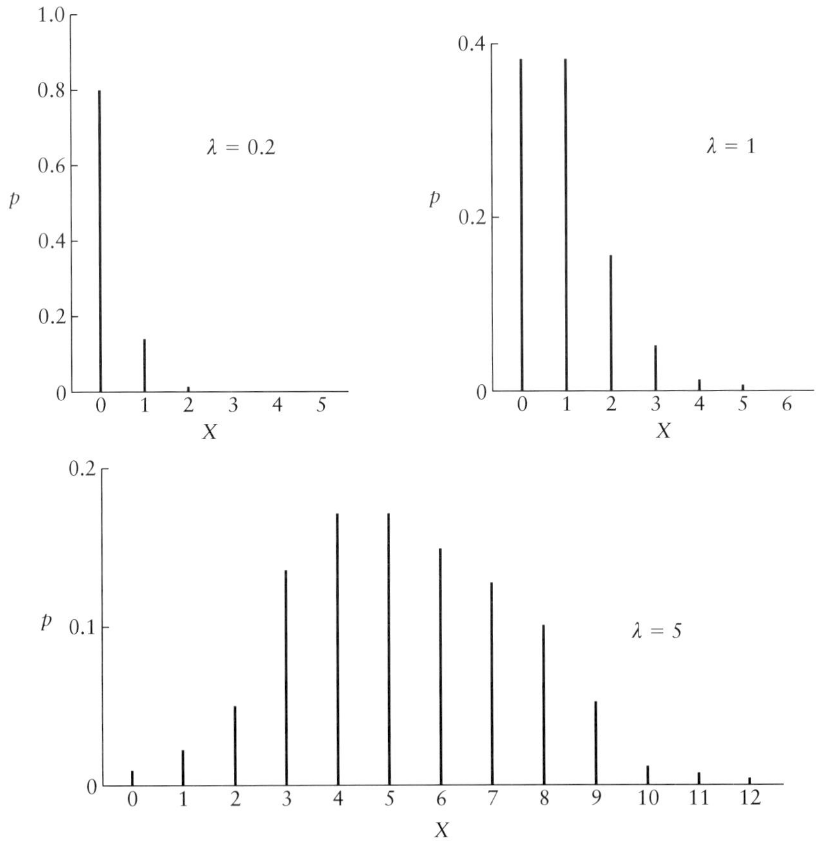

Figure 2.2 *The shape of the Poisson distribution for (a) $\lambda = 0.2$ (b) $\lambda = 1$ (c) $\lambda = 5$*

In the previous few pages, you have been introduced to the Poisson distribution as a very good approximation to the binomial distribution. However, it is much more significant than that. It is a distribution in its own right which can be used to model situations which are clearly not binomial, where you cannot state the number of trials or the probabilities of success and failure.

CONDITIONS UNDER WHICH THE POISSON DISTRIBUTION MAY BE USED

The Poisson distribution is generally thought of as the probability distribution for the number of occurrences of a *rare event.*

The conditions under which it may be used are as follows:

1. As an approximation to the binomial distribution
The Poisson distribution may be used as an approximation to the binomial distribution, $B(n, p)$, when

(a) n is large
(b) p is small (and so the event is rare).

In addition you would not usually want to use the Poisson distribution for large values of np, so in practice (but not in theory) a third condition may be applied:

(c) np is not so large that the Poisson probabilities are more difficult to calculate than their binomial counterparts.

In the case of the disease, n was large at 60 000, p small at $\frac{1}{40\,000}$. In addition np, 1.5, was not large.

It is also necessary that the trials are random and independent; otherwise the distribution to be approximated would not be binomial in the first place.

2. As a distribution in its own right
There are many situations in which events happen singly and the mean number of occurrences is known (or can easily be found) but it is not possible, or even meaningful, to give values to the number of trials, n, or the probability of success, p. You can, for example, find the mean number of goals per team in league football matches, or the mean number of telephone calls received per minute at an exchange, but you cannot say how many goals are not scored or telephone calls not received. The concept of a trial, with success or failure as possible outcomes, is not appropriate to these situations. They may, however, be modelled by the Poisson distribution, provided that

(a) the occurrences are random
(b) the occurrences are independent
(c) there is a known, constant overall mean rate for the occurrences.

It would be unusual to use the Poisson distribution with a population mean greater than about 20, because easier methods exist.

EXAMPLE 2.1

Nicky buys a lottery ticket every week for 50 weeks. She reckons that the probability of any individual ticket winning a prize is 0.01.

(a) Write down the distribution of X, the number of winning tickets out of 50.
(b) Give a suitable approximating distribution.
(c) Use your approximation to calculate the probability that she wins exactly two prizes in 50 weeks.

Solution **(a)** $X \sim B(50, 0.01)$

Notation: B(50, 0.01) indicates a binomial distribution with $n = 50$ and $p = 0.01$

(b) Since n is large (50), p is small (0.01) and np is not too big, X may be approximated by a Poisson distribution with parameter $\lambda = np = 0.5$. Thus the approximation is $X \sim Po(0.5)$.

Notation: Po(0.5) indicates a Poisson distribution with parameter $\lambda = 0.5$

(c) Using Po(0.5)

$$P(X = 2) = e^{-0.5} \frac{0.5^2}{2!}$$
$$= 0.0758$$

EXAMPLE 2.2

The number of defects in a wire cable can be modelled by the Poisson distribution with a mean of four defects per kilometre.

What is the probability that a single kilometre of wire will have exactly two defects?

Solution Let X = the number of defects per kilometre. Using $X \sim Po(4)$,

$$P(X = 2) = e^{-4} \frac{4^2}{2!}$$
$$= 0.147$$

EXERCISE 2A

1 The random variable X is binomially distributed with $n = 80$ and $p = 0.05$. Write down a suitable approximating distribution and use it to find the probability that $X = 3$.

2 The random variable $X \sim B(60, 0.04)$. Write down a suitable approximating distribution and use it to find the probability that $X = 2$.

3 The random variable X is Poisson distributed with $\lambda = 5$. Find the probability that

(a) $X = 3$ **(b)** $X = 4$ **(c)** $X = 5$

4 The random variable $X \sim \text{Po}(2.8)$. Find the probability that

(a) $X = 2$ **(b)** $X = 3$ **(c)** $X = 4$

5 The number of wombats which are killed on a particular stretch of road in Australia in any one day can be modelled by a Po(0.4) random variable. Calculate the probability that exactly two wombats are killed on a given day on this stretch of road.

[Cambridge]

THE MEAN AND VARIANCE OF Po(λ)

It may be shown that the mean and variance of Po(λ) are both equal to the parameter λ. You do not need to know the theory of the derivation of these results, but you may need to use them to justify a Poisson model.

EXAMPLE 2.3

Peter believes that a Poisson distribution may model the number of goals scored by his favourite football team. He analyses a large set of past results, finding a mean of 1.2 and standard deviation 1.1 goals per match.

(a) Calculate the variance from Peter's data, and explain how its value lends support to the Poisson model.

(b) Using a Poisson distribution with mean 1.2, find the probability that in their next match Peter's favourite team will score

(i) no goals **(ii)** one goal **(iii)** more than one goal.

Solution

(a) Variance $= sd^2$
$= 1.1^2$
$= 1.21$

Since $1.21 \approx 1.2$, the variance $\approx$ the mean, which supports the Poisson model.

(b) Using Po(1.2)

(i) $P(X = 0) = \dfrac{e^{-1.2} \times 1.2^0}{0!} = 0.3012$

(ii) $P(X = 1) = \dfrac{e^{-1.2} \times 1.2^1}{1!} = 0.3614$

(iii) $P(X > 1) = 1 - [P(X = 0) + P(X = 1)]$

$= 1 - (0.3012 + 0.3614)$

$= 1 - 0.6626$

$= 0.3374$

Calculating poisson distribution probabilities

There are two ways of reducing the often tedious calculations involved in finding Poisson probabilities, cutting down on the amount of work, and so on the time you take.

Recurrence relations

Recurrence relations allow you to use the term you have obtained to work out the next one. For the Poisson distribution with parameter λ,

$P(X = 0) = e^{-\lambda}$ You must use your calculator to find this term.

$P(X = 1) = e^{-\lambda}\lambda = \lambda P(X = 0)$ Multiply the previous term by λ.

$$P(X = 2) = e^{-\lambda}\frac{\lambda^2}{2!} = \frac{\lambda}{2}P(X = 1)$$

Multiply the previous term by $\frac{\lambda}{2}$.

$$P(X = 3) = e^{-\lambda}\frac{\lambda^3}{3!} = \frac{\lambda}{3}P(X = 2)$$

Multiply the previous term by $\frac{\lambda}{3}$.

$$P(X = 4) = e^{-\lambda}\frac{\lambda^4}{4!} = \frac{\lambda}{4}P(X = 3)$$

Multiply the previous term by $\frac{\lambda}{4}$.

In general, you can find $P(X = r)$ by multiplying your previous probability, $P(X = r - 1)$, by $\frac{\lambda}{r}$. You would expect to hold the latest value on your calculator and keep a running total in the memory.

Setting this out on paper with $\lambda = 1.5$, for example, gives these figures:

r	Conversion	$P(X = r)$	Running total, $P(X \leqslant r)$
0		0.223 130	0.223 130
	$\times 1.5$		
1		0.334 695	0.557 825
	$\times \frac{1.5}{2}$		
2		0.251 021	0.808 846
	$\times \frac{1.5}{3}$		
3		0.125 511	0.934 357
	$\times \frac{1.5}{4}$		
4		0.047 067	0.981 424

EXAMPLE 2.4

$X \sim \text{Po}(3.5)$. Find the probability that X is less than or equal to 3.

Solution Using Po(3.5)

$P(X = 0) = e^{-3.5} = 0.0302$

$P(X = 1) = [ANS] \times 3.5 = 0.1057$

$P(X = 2) = [ANS] \times \dfrac{3.5}{2} = 0.1850$

$P(X = 3) = [ANS] \times \dfrac{3.5}{3} = 0.2158$

Here [*ANS*] refers to the answer to the previous calculation. Your calculator probably has such a key on it

Summing these probabilities,

$$P(X \leqslant 3) = 0.0302 + 0.1057 + 0.1850 + 0.2158 = 0.5367$$

CUMULATIVE POISSON TABLES

You can also find cumulative probabilities directly, using cumulative tables:

POISSON CUMULATIVE DISTRIBUTION FUNCTION

The tabulated value is $P(X \leqslant x)$, where X has a Poisson distribution with parameter λ.

$\lambda =$	0.5	1.0	1.5	2.0	2.5	3.0	3.5	4.0	4.5	5.0
$x = 0$	0.6065	0.3679	0.2231	0.1353	0.0821	0.0498	0.0302	0.0183	0.0111	0.0067
1	0.9098	0.7358	0.5578	0.4060	0.2873	0.1991	0.1359	0.0916	0.0611	0.0404
2	0.9856	0.9197	0.8088	0.6767	0.5438	0.4232	0.3208	0.2381	0.1736	0.1247
3	0.9982	0.9810	0.9344	0.8571	0.7576	0.6472	0.5366	0.4335	0.3423	0.2650
4	0.9998	0.9963	0.9814	0.9473	0.8912	0.8153	0.7254	0.6288	0.5321	0.4405
5	1.0000	0.9994	0.9955	0.9834	0.9580	0.9161	0.8576	0.7851	0.7029	0.6160
6	1.0000	0.9999	0.9991	0.9955	0.9858	0.9665	0.9347	0.8893	0.8311	[illegible]
7	1.0000	[illegible]								

FIGURE 2.3

Note Not all Poisson distributions are tabulated, so you must ensure that you understand both methods of solving Example 2.4.

EXERCISE 2B

1 If $W \sim \text{Po}(3)$, calculate

(a) $P(W = 0)$ **(b)** $P(W = 1)$ **(c)** $P(W = 2)$

(d) $P(W \leqslant 2)$ **(e)** $P(W > 2)$.

2 If $X \sim \text{Po}(4.5)$, calculate **(a)** $P(X \leqslant 3)$ **(b)** $P(X > 3)$.

3 The number of cars passing a house in a residential road between 10 am and 11 am on a weekday is a random variable, X. Give a condition under which X may be modelled by a Poisson distribution.

Suppose that $X \sim \text{Po}(2.5)$. Calculate $\text{P}(X \geqslant 5)$.

[Cambridge, adapted]

4 The number of night calls to a fire station in a small town can be modelled by a Poisson distribution with mean 4.2 per night. Find the probability that on a particular night there will be three or more calls to the fire station.

State what needs to be assumed about the calls to the fire station in order to justify a Poisson model.

[Cambridge]

5 During the 1998–99 season Homerton F.C. scored the following numbers of goals in their league matches:

Goals	0	1	2	3	4	5	6
Matches	7	11	10	8	3	1	2

(a) How many league matches did Homerton play that season?
(b) Calculate the mean and variance of the number of goals per match.
(c) Using the mean you calculated in part **(a)** as the parameter in the Poisson distribution, calculate the probabilities of $0, 1, \ldots, 5, 6$ or more goals in any match.
(d) In how many matches would you expect Homerton to have scored $0, 1, 2, \ldots, 6$ or more goals, according to the Poisson model?
(e) State, with reasons, whether you consider this to be a good model.

6 In a country the mean number of deaths per year from lightning strike is 2.2.
(a) Find the probabilities of 0, 1, 2 and more than 2 deaths from lightning strike in any particular year.

In a neighbouring country, it is found that one year in twenty nobody dies from lightning strike.
(b) What is the mean number of deaths per year in that country from lightning strike?

7 A manufacturer of rifle ammunition tests a large consignment for accuracy by firing 500 batches, each of 20 rounds, from a fixed rifle at a target. Those rounds that fall outside a marked circle on the target are classified as *misses*. For each batch of 20 rounds the number of misses is counted.

Misses, X	0	1	2	3	4	5	6–20
Frequency	230	189	65	15	0	1	0

(a) Estimate the mean number of misses per batch.

(b) Use your mean to estimate the probability of a batch producing 0, 1, 2, 3, 4 and 5 misses using the Poisson distribution as a model.

(c) Use your answers to part **(b)** to estimate expected frequencies of 0, 1, 2, 3, 4 and 5 misses per batch in 500 batches and compare your answers with those actually found.

(d) Do you think the Poisson distribution is a good model for this situation?

8 A survey in a town's primary schools has indicated that 5% of the pupils have severe difficulties with reading. If the primary school pupils were allocated to the secondary schools at random, estimate the probability that a secondary school with an intake of 200 pupils will receive

(a) no more than 8 pupils with severe reading difficulties

(b) more than 20 pupils with severe reading difficulties.

9 Fanfold paper for computer printers is made by putting perforations every 30 cm in a continuous roll of paper. A box of fanfold paper contains 2000 sheets. State the length of the continuous rolls from which the box of paper is produced.

The manufacturers claim that faults occur at random and at an average rate of 1 per 240 metres of paper. State an appropriate distribution for the number of faults per box of paper. Find the probability that a box of paper has no faults and also the probability that it has more than 4 faults.

Two copies of a report which runs to 100 sheets per copy are printed on this sort of paper. Find the probability that there are no faults in either copy of the report and also the probability that just one copy is faulty.

[MEI]

10 A firm investigated the number of employees suffering injuries whilst at work. The results recorded below were obtained for a 52-week period:

Number of employees injured in a week	0	1	2	3	4 or more
Number of weeks	31	17	3	1	0

Give reasons why one might expect this distribution to approximate to a Poisson distribution. Evaluate the mean and variance of the data and explain why this gives further evidence in favour of a Poisson distribution.

Using the calculated value of the mean, find the theoretical frequencies of a Poisson distribution for the number of weeks in which 0, 1, 2, 3, 4 or more employees were injured.

SUMMING AND SCALING POISSON DISTRIBUTIONS

It is often the case that you wish to rescale a Poisson distribution. For example, if the number of road accidents per day in a metropolitan area is Poisson distributed with parameter 10.1 then the number per week (of 7 days) will be Poisson distributed with parameter 70.7; you just multiply the parameter by 7.

This is a special case of a more general principle:

if $X \sim \text{Po}(\lambda_1)$ and $Y \sim \text{Po}(\lambda_2)$ then $X + Y \sim \text{Po}(\lambda_1 + \lambda_2)$.

EXAMPLE 2.5

A small telephone switchboard receives, on average, one call every 30 seconds during the mid-afternoon. Find the probability that, during one particular mid-afternoon, it receives

(a) exactly one call in a given period of 30 seconds
(b) no calls in a given period of one minute
(c) at least eight calls in a given period of five minutes.

Solution

(a) Let X be the number of calls received in a 30-second period.
$X \sim \text{Po}(1)$

$$P(X = 1) = e^{-1}\frac{1^1}{1!} = 0.3679$$

(b) Let V be the number of calls received in a 1-minute period.
$V \sim \text{Po}(2)$

$$P(V = 0) = e^{-2}\frac{2^0}{0!} = 0.1353$$

(c) Let W be the number of calls received in a 5-minute period.
$W \sim \text{Po}(10)$

$$P(W \geqslant 8) = 1 - P(W \leqslant 7)$$

$$= 1 - 0.2202 \quad \text{using cumulative tables of Po(10)}$$

$$= 0.7798$$

EXAMPLE 2.6

A rare disease causes the death, on average, of 3.7 people per year in England, 0.8 in Scotland and 0.5 in Wales. As far as is known the disease strikes at random and cases are independent of one another.

What is the probability of 7 or more deaths from the disease on the British mainland (i.e. England, Scotland and Wales together) in any year?

Solution Notice first that:

(a) P(7 or more deaths) = 1 – P(6 or fewer deaths)

(b) each of the three distributions fulfils the conditions for it to be modelled by the Poisson distribution.

You could find the probability of 6 or fewer deaths by listing all the different ways they could be allocated between the three countries (e.g. England 3, Scotland 1, Wales 2) and working out the probabilities of all of them, using the three individual Poisson distributions. That would be very time-consuming indeed.

It is much easier to add the three distributions together and treat the result as a single Poisson distribution.

The overall mean is given by	3.7	+	0.8	+	0.5	=	5.0
	England		Scotland		Wales		Total

giving an overall distribution of Po(5.0).

The probability of 6 or fewer deaths is then found from cumulative Poisson probability tables to be 0.7622.

So the probability of 7 or more deaths is given by 1 – 0.7622 = 0.2378

Note You may only add Poisson distributions in this way if they are independent of each other.

EXAMPLE 2.7

On a lonely Highland road in Scotland, cars are observed passing at the rate of 6 per day, and lorries at the rate of 2 per day. On the road is an old cattle grid which will soon need repair. The local works department decide that if the probability of more than 15 vehicles per day passing is less than 1% then the repairs to the cattle grid can wait until next spring, otherwise it will have to be repaired before the winter.

When will the cattle grid have to be repaired?

Solution Let C be the number of cars per day, L be the number of lorries per day, and V be the number of vehicles per day.

$$V = L + C$$

Assuming that a car or a lorry passing along the road is a random event and the two are independent:

$$C \sim \text{Po}(6), \quad L \sim \text{Po}(2)$$

and so $V \sim \text{Po}(6 + 2)$

i.e. $V \sim \text{Po}(8)$

From cumulative Poisson probability tables $P(V \leqslant 15) = 0.9918$.

The required probability is $P(V > 15) = 1 - P(V \leqslant 15)$

$= 1 - 0.9918$

$= 0.0082$

This is just less than 1% and so the repairs are left until spring.

EXERCISE 2C

1 Betty drives along a 50-mile stretch of motorway 5 days a week 50 weeks a year. She takes no notice of the 70 mph speed limit and, when the traffic allows, travels between 95 and 105 mph. From time to time she is caught by the police and fined but she estimates the probability of this happening on any day is $\frac{1}{300}$. If she gets caught three times within three years she will be disqualified from driving. Use Betty's estimates of probability to answer the following questions.

(a) What is the probability of her being caught exactly once in any year?

(b) What is the probability of her being caught less than three times in three years?

(c) What is the probability of her being caught exactly three times in three years?

Betty is in fact caught one day and decides to be somewhat cautious, reducing her normal speed to between 85 and 95 mph. She believes this will reduce the probability of her being caught to $\frac{1}{500}$.

(d) What is the probability that she is caught less than twice in the next three years?

2 Motorists in a particular part of the Highlands of Scotland have a choice between a direct route and a one-way scenic detour. It is known that on average one in forty of the cars on the road will take the scenic detour. The road engineer wishes to do some repairs on the scenic detour. He chooses a time when he expects 100 cars an hour to pass along the road.

Find the probability that, in any one hour,

(a) no cars **(b)** at most 4 cars

will turn on to the scenic detour.

(c) Between 10.30 am and 11.00 am it will be necessary to block the road completely. What is the probability that no car will be delayed?

3 A sociologist claims that only 3% of all suitably qualified students from inner city schools go on to university. Use his claim and the Poisson approximation to the binomial distribution to estimate the probability that in a randomly chosen group of 200 such students

(a) exactly five go to university

(b) more than five go to university.

(c) If there is at most a 5% chance that more than n of the 200 students go to university, find the lowest possible value of n.

Another group of 100 students is also chosen. Find the probability that

(d) exactly five of each group go to university

(e) exactly ten of all the chosen students go to university.

[MEI]

4 At the hot drinks counter in a cafeteria both tea and coffee are sold. The number of cups of coffee sold per minute may be assumed to be a Poisson variable with mean 1.5 and the number of cups of tea sold per minute may be assumed to be an independent Poisson variable with mean 0.5.

(a) Calculate the probability that in a given one-minute period exactly one cup of tea and one cup of coffee are sold.

(b) Calculate the probability that in a given three-minute period fewer than five drinks altogether are sold.

(c) In a given one-minute period exactly three drinks are sold. Calculate the probability that these are all cups of coffee.

[Cambridge]

5 The numbers of lorry drivers and car drivers visiting an all-night transport cafe between 2 am and 3 am on a Sunday morning have independent Poisson distributions with means 5.1 and 3.6 respectively. Find the probabilities that, between 2 am and 3 am on any Sunday,

(a) exactly five lorry drivers visit the cafe

(b) at least one car driver visits the cafe

(c) exactly five lorry drivers and exactly two car drivers visit the cafe.

By using the distribution of the *total* number of drivers visiting the cafe, find the probability that exactly seven drivers visit the cafe between 2 am and 3 am on any Sunday. Given that exactly seven drivers visit the cafe between 2 am and 3 am on one Sunday, find the probability that exactly five of them are driving lorries.

[MEI]

6 A garage uses a particular spare part at an average rate of five per week. Assuming that usage of this spare part follows a Poisson distribution, find the probability that

(a) exactly five are used in a particular week

(b) at least five are used in a particular week

(c) exactly ten are used in a two-week period

(d) at least ten are used in a two-week period

(e) exactly five are used in each of two successive weeks.

If stocks are replenished weekly, determine the number of spare parts which should be in stock at the beginning of each week to ensure that on average the stock will be insufficient on no more than one week in a 52-week year.

7 Small hard particles are found in the molten glass from which glass bottles are made. On average, 15 particles are found per 100 kg of molten glass. If a bottle contains one or more such particles it has to be discarded.

Suppose bottles of mass 1 kg are made. It is required to estimate the percentage of bottles that have to be discarded. Criticise the following 'answer': *Since the material for 100 bottles contains 15 particles, approximately 15% will have to be discarded.*

Making suitable assumptions, which should be stated, develop a correct argument using a Poisson model, and find the percentage of faulty 1 kg bottles to three significant figures.

Show that about 3.7% of bottles of mass 0.25 kg are faulty.

[MEI]

8 Weak spots occur at random in the manufacture of a certain cable at an average rate of 1 per 100 metres. If X represents the number of weak spots in 100 metres of cable, write down the distribution of X.

Lengths of this cable are wound on to drums. Each drum carries 50 metres of cable. Find the probability that a drum will have three or more weak spots.

A contractor buys five such drums. Find the probability that two have just one weak spot each and the other three have none.

[AEB]

9 A crockery manufacturer tests dinner plates by taking a large sample from each day's production. When all the machinery is set correctly, there are on average two faulty plates per sample but this number rises if any part of the process is incorrectly set.

When the machine is working correctly, what is the probability that a test will result in

(a) no faulty plates

(b) at most four faulty plates?

The company reset the machinery (a process which is expensive in time) if there are five or more faulty plates in a sample.

(c) What is the probability that the machinery is reset unnecessarily?

The company decide to change the basis on which they make the decision to reset their machinery so that they will now do so if the total number of faulty plates in three consecutive samples is at least f.

(d) Find the smallest value of f which gives a probability of less than 1% that the machinery will be reset unnecessarily.

(e) Given your value of f, find the probability that a set of three consecutive samples fails to indicate that the machinery needs resetting when the mean number of faults per sample has in fact risen to 3.0.

10 A petrol station has service areas on both sides of a motorway, one to serve east-bound traffic and the other for west-bound traffic. The number of east-bound vehicles arriving at the station in one minute has a Poisson distribution with mean 0.9, and the number of west-bound vehicles arriving in one minute has a Poisson distribution with mean 1.6, the two distributions being independent.

(a) Find the probability that in a one-minute period

(i) no vehicles arrive

(ii) more than two vehicles arrive at this petrol station,

giving your answers correct to three places of decimals.

Given that in a particular one-minute period three vehicles arrive, find

(b) the probability that they are all from the same direction

(c) the most likely combination of east-bound and west-bound vehicles.

[Cambridge]

EXERCISE 2D Examination-style questions

1 350 raisins are put into a mixture which is well stirred and made into 100 small buns. Estimate how many of these buns will

(a) be without raisins

(b) contain five or more raisins.

In a second batch of 100 buns, exactly one has no raisins in it.

(c) Estimate the total number of raisins in the second mixture.

2 In Abbotson town centre, the number of incidents of criminal damage reported to the police averages two per week.

(a) Explain why the Poisson distribution might be thought to be a suitable model for the number of incidents of criminal damage reported per week.

(b) Find the probabilities of the following events, according to the Poisson model.

(i) Exactly two incidents are reported in a week.

(ii) Two consecutive weeks are incident-free.

(iii) More than ten incidents are reported in a period of four weeks.

[MEI]

3 The probability that I dial a wrong number when making a telephone call is 0.015. In a typical week I will make 50 telephone calls. Using a Poisson approximation to a binomial model find, correct to two decimal places, the probability that in such a week

(a) I dial no wrong numbers

(b) I dial more than two wrong numbers.

[Cambridge, part]

4 A Christmas draw aims to sell 5000 tickets, 50 of which will each win a prize.

(a) A syndicate buys 200 tickets. Let X represent the number of tickets that win a prize.

(i) Justify the use of the Poisson approximation for the distribution of X.

(ii) Calculate $P(X \leqslant 3)$.

(b) Calculate how many tickets should be bought in order for there to be a 90% probability of winning at least one prize.

[Cambridge]

5 A count was made of the number of red blood corpuscles in each of the 64 compartments of a haemocytometer with the following results:

Number of corpuscles	2	3	4	5	6	7	8	9	10	11	12	13	14
Frequency	1	5	4	9	10	10	8	6	4	3	2	1	1

Estimate the mean and variance of the number of red blood corpuscles per compartment. Explain how the values you have obtained support the view that these data are a sample from a Poisson population.

Write down an expression for the theoretical frequency with which compartments containing five red blood corpuscles should be found, assuming this to be obtained from a Poisson population with mean 7. Evaluate this frequency to two decimal places.

[MEI]

6 At a busy intersection of roads, accidents requiring the summoning of an ambulance occur with a frequency, on average, of 1.5 per week. These accidents occur randomly, so that it may be assumed that they follow a Poisson distribution.

(a) Calculate the probability that there will not be an accident in a given week.

(b) Calculate the smallest integer n such that the probability of more than n accidents in a week is less than 0.02.

(c) Calculate the probability that there will not be an accident in a given fortnight.

(d) Calculate the largest integer k such that the probability that there will not be an accident in k successive weeks is greater than 0.0001.

[AEB, adapted]

7 A ferry takes cars and small vans on a short journey from an island to the mainland. On a representative sample of weekday mornings, the numbers of vehicles, X, on the 8 am sailing were as follows:

20	24	24	22	23	21	20	22	23	22
21	21	22	21	23	22	20	22	20	24

(a) Show that X does not have a Poisson distribution.

In fact 20 of the vehicles belong to commuters who use that sailing of the ferry every weekday morning. The random variable Y is the number of vehicles other than those 20 who are using the ferry.

(b) Investigate whether Y may reasonably be modelled by a Poisson distribution.

The ferry can take 25 vehicles on any journey.

(c) On what proportion of days would you expect at least one vehicle to be unable to travel on this particular sailing of the ferry because there was no room left and so have to wait for the next one?

8 A traffic survey is being undertaken on a main road to determine whether or not a pedestrian crossing should be installed. On five successive days, from Monday to Friday, the hour between 8 am and 9 am was split up into 30-second intervals and the number of vehicles passing a certain point in each of these intervals was recorded.

The random variable X represents the number of cars travelling *from* the town centre per 30-second interval. For the 600 observations the mean and variance were 3.1 and 3.27 respectively.

(a) Explain why X might be modelled by a Poisson distribution.

(b) Using the sample mean as an estimate for the Poisson parameter, calculate the probability of recording exactly 3 vehicles travelling from the town centre in a 30-second interval.

(c) Calculate the probability of recording at least 6 vehicles travelling from the town centre in a 60-second interval.

[MEI]

9 A car hire firm has three cars, which it hires out on a daily basis. The number of cars demanded per day follows a Poisson distribution with mean 2.1.

(a) Find the probability that exactly two cars are hired out on any one day.

(b) Find the probability that all cars are in use on any one day.

(c) Find the probability that all cars are in use on exactly 3 days of a 5-day week.

(d) Find the probability that exactly 10 cars are demanded in a 5-day week. Explain whether or not such a demand could always be met.

(e) It costs the firm £20 a day to run each car, whether it is hired out or not. The daily hire charge per car is £50. Find the expected daily profit.

[MEI]

10 The following table gives the number f_r of each of 519 equal time intervals in which r radioactive atoms decayed.

Number of decays, r	0	1	2	3	4	5	6	7	8	$\geqslant 9$
Observed number of intervals, f_r	11	41	73	105	107	82	55	28	9	8

Estimate the mean and variance of r.

Suggest, with justification, a theoretical distribution from which the data could be a random sample. Hence calculate expected values of f_r and comment briefly on the agreement between these and the observed values.

In the experiment each time interval was of length 7.5 s. In a further experiment, 1000 time intervals each of length 5 s are to be examined. Estimate the number of these intervals within which no atoms will decay.

Historical note

Simeon Poisson was born in Pithiviers in France in 1781. Under family pressure he began to study medicine but after some time gave it up for his real interest, mathematics. For the rest of his life Poisson lived and worked as a mathematician in Paris. His contribution to the subject spanned a broad range of topics in both pure and applied mathematics, including integration, electricity and magnetism and planetary orbits as well as statistics. He was the author of between 300 and 400 publications and originally derived the Poisson distribution as an approximation to the binomial distribution.

When he was a small boy, Poisson had his hands tied by his nanny who then hung him from a hook on the wall so that he could not get into trouble while she went out. In later life he devoted a lot of time to studying the motion of a pendulum and claimed that this interest derived from his childhood experience swinging against the wall.

KEY POINTS

1 THE POISSON PROBABILITY DISTRIBUTION

- If $X \sim$ Poisson (λ) the parameter $\lambda > 0$.

$$P(X = r) = e^{-\lambda} \frac{\lambda^r}{r!} \quad r \geqslant 0, r \text{ is an integer}$$

$$\text{Mean}(X) = \lambda$$

$$\text{Var}(X) = \lambda$$

2 CONDITIONS UNDER WHICH THE POISSON DISTRIBUTION MAY BE USED

- The Poisson distribution is generally thought of as the probability distribution for the number of occurrences of a *rare event.*

- *As a distribution in its own right*
  Situations in which the mean number of occurrences is known (or can easily be found) but it is not possible, or even meaningful, to give values to n or p may be modelled using the Poisson distribution provided that the occurrences are

  — random
  — independent.

- *As an approximation to the binomial distribution*
  The Poisson distribution may be used as an approximation to the binomial distribution, $B(n, p)$, when

  — n is large
  — p is small (and so the event is rare)
  — np is not large.

It would be unusual to use the Poisson distribution with parameter, λ, greater than about 20.

3 THE SUM OF TWO POISSON DISTRIBUTIONS

If $X \sim Po(\lambda_1)$, $Y \sim Po(\lambda_2)$ and X and Y are independent
$X + Y \sim Po(\lambda_1 + \lambda_2)$.

Chapter three

CONTINUOUS RANDOM VARIABLE

I can't help it, the idea of the infinite torments me.

Alfred de Musset

Consider the heights of waves in the sea at a certain time of year. Suppose that all the waves are between 0.5 m and 12 m high, with most of them being between 3 m and 8 m high. The distribution of the heights of the waves might be modelled by a graph like that in figure 3.1, which was drawn from data collected by an offshore weather ship.

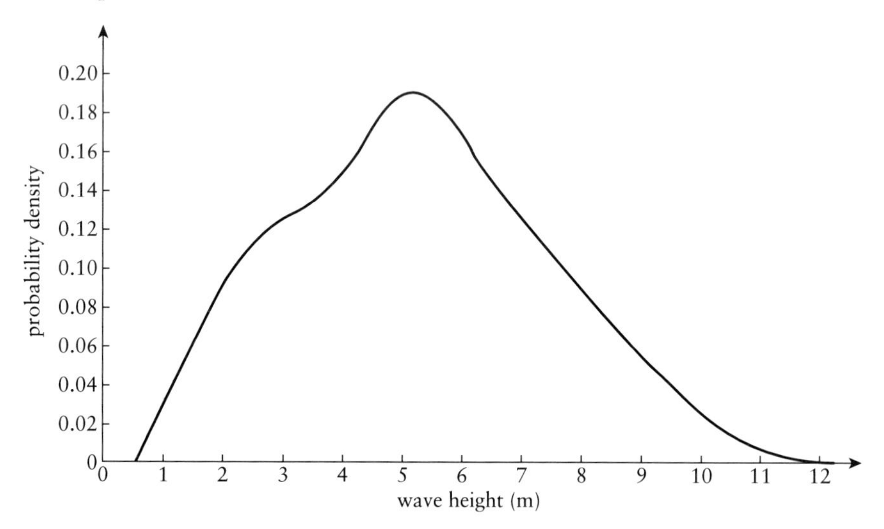

FIGURE 3.1

Figure 3.1 shows the *relative likelihoods* of various wave sizes. For example, it shows that waves between 4 m and 5 m high occur more often than waves between 2 m and 3 m high, since the curve is higher in that region.

Note that it is meaningless to talk about the probability of a wave being *exactly* 4 m high, as such a result is impossible. If you took a wave which seemed to be 4 m high, and you were able to measure it to enough decimal places, sooner or later you would find it was 4.000 13 or 3.999 96 or whatever, i.e. it will not be exactly 4 m. This is because the height of a wave is a *continuous random variable*, i.e. it can take any one of an infinite number of possible values.

It is, however, perfectly all right to discuss the probability of the wave being *between* 4 m and 5 m high. This situation corresponds to the shaded area shown in figure 3.2 below.

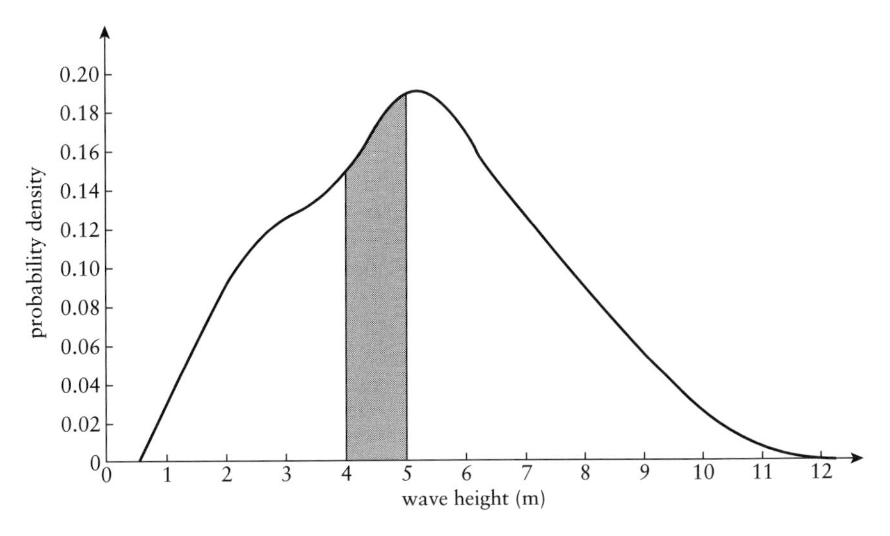

FIGURE 3.2

Look at the vertical scale on figures 3.1 and 3.2, and you will see decimals 0.10, 0.12, etc. These values are labelled as the *probability density*.

The reason that the y axis has been labelled using this scale is to force the total area under the curve to be exactly 1 unit. This means that the area between 4 and 5, shaded in figure 3.2, gives the probability directly, without the need to divide by anything.

PROBABILITY DENSITY FUNCTION

In the wave height example the curve was determined experimentally, using equipment on board the offshore weather ship. The curve is continuous because the random variable, the wave height, is continuous and not discrete. The possible heights of waves are not restricted to particular steps (say every $\frac{1}{2}$ metre), but may take any value within a range.

A function represented by a curve of this type is called a *probability density function*, often abbreviated to p.d.f. The probability density function of a continuous random variable, X, is usually denoted by $f(x)$. If $f(x)$ is a p.d.f. it follows that:

- $f(x) \geqslant 0$ for all x — You cannot have negative probabilities.
- $\displaystyle\int_{\text{All values of } x} f(x)\,dx = 1$ — The total area under the curve is 1.

For a continuous random variable with probability density function f(x), the probability that X lies in the interval $[a, b]$ is given by

$$P(a \leqslant X \leqslant b) = \int_a^b f(x)\,dx.$$

You will see that in this case the probability density function has quite a complicated curve and so it is not possible to find a simple algebraic expression with which to model it.

EXAMPLE 3.1

An insurance company are thinking of fitting an electronic security system inside head office. They have been told by manufacturers that the lifetime, X years, of the system they have in mind has the p.d.f.:

$$f(x) = \frac{3x(20 - x)}{4000} \quad \text{for } 0 \leqslant x \leqslant 20,$$

and $\quad f(x) = 0 \quad$ otherwise.

(a) Show that the manufacturers' statement is consistent with f(x) being a probability density function.

(b) Find the probability that:

(i) it fails in the first year

(ii) it lasts 10 years but then fails in the next year.

Solution

(a) The condition $f(x) \geqslant 0$ for all values of x between 0 and 20 is satisfied, as shown by the graph of f(x), figure 3.3.

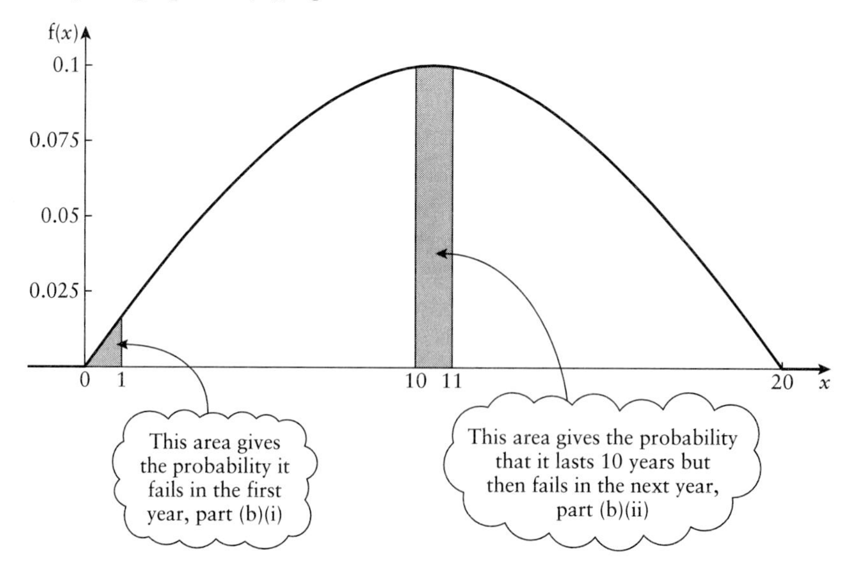

FIGURE 3.3

The other condition is that the area under the curve is 1.

$$\text{Area} = \int_{-\infty}^{\infty} \mathrm{f}(x)\,\mathrm{d}x = \int_0^{20} \frac{3x(20-x)}{4000}\,\mathrm{d}x$$

$$= \frac{3}{4000}\int_0^{20} (20x - x^2)\,\mathrm{d}x$$

$$= \frac{3}{4000}\left[10x^2 - \frac{x^3}{3}\right]_0^{20}$$

$$= \frac{3}{4000}\left[10 \times 20^2 - \frac{20^3}{3}\right]$$

$= 1$, as required.

(b) (i) *It fails in the first year.*

This is given by $\mathrm{P}(X < 1) = \int_0^{1} \frac{3x(20-x)}{4000}\,\mathrm{d}x$

$$= \frac{3}{4000}\int_0^{1} (20x - x^2)\,\mathrm{d}x$$

$$= \frac{3}{4000}\left[10x^2 - \frac{x^3}{3}\right]_0^{1}$$

$$= \frac{3}{4000}\left(10 \times 1^2 - \frac{1^3}{3}\right)$$

$= 0.007\,25$

(ii) *It fails in the 11th year.*

This is given by $\mathrm{P}(10 \leqslant X < 11)$

$$= \int_{10}^{11} \frac{3x(20-x)}{4000}\,\mathrm{d}x$$

$$= \frac{3}{4000}\left[10x^2 - \frac{1}{3}x^3\right]_{10}^{11}$$

$$= \frac{3}{4000}\left(10 \times 11^2 - \frac{1}{3} \times 11^3\right) - \frac{3}{4000}\left(10 \times 10^2 - \frac{1}{3} \times 10^3\right)$$

$= 0.074\,75$

EXAMPLE 3.2

The continuous random variable X represents the amount of sunshine in hours between noon and 4 pm at a skiing resort in the high season. The probability density function, $f(x)$, of X is modelled by

$$f(x) = \begin{cases} kx^2 & \text{for } 0 \leqslant x \leqslant 4 \\ 0 & \text{otherwise.} \end{cases}$$

(a) Find the value of k.

(b) Find the probability that on a particular day in the high season there is more than two hours of sunshine between noon and 4 pm.

Solution

(a) To find the value of k you must use the fact that the area under the graph of $f(x)$ is equal to 1.

$$\int_{-\infty}^{\infty} f(x)\,dx = \int_0^4 kx^2\,dx = 1$$

Therefore
$$\left[\frac{kx^3}{3}\right]_0^4 = 1$$

$$\frac{64k}{3} = 1$$

So
$$k = \frac{3}{64}$$

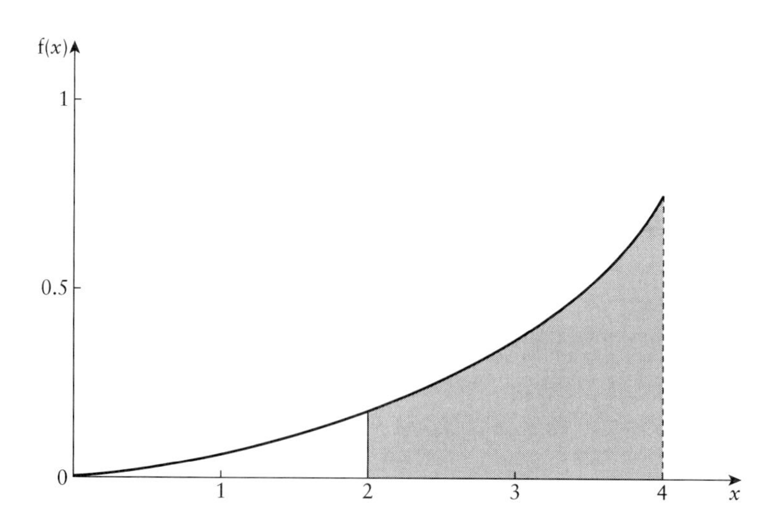

Figure 3.4

(b) The probability of more than 2 hours of sunshine is given by

$$P(X > 2) = \int_2^{\infty} f(x)\,dx = \int_2^4 \frac{3x^2}{64}\,dx$$
$$= \left[\frac{x^3}{64}\right]_2^4$$
$$= \frac{64 - 8}{64}$$
$$= \frac{56}{64}$$
$$= 0.875$$

EXAMPLE 3.3

The number of hours Darren spends each day working in his garden is modelled by the continuous random variable X, with p.d.f. $f(x)$ defined by

$$f(x) = \begin{cases} kx & \text{for } 0 \leqslant x < 3 \\ k(6 - x) & \text{for } 3 \leqslant x \leqslant 6 \\ 0 & \text{otherwise.} \end{cases}$$

(a) Find the value of k.

(b) Sketch the graph of $f(x)$.

(c) Find the probability that Darren will work between 2 and 5 hours in his garden on a randomly selected day.

Solution **(a)** To find the value of k you must use the fact that the area under the graph of $f(x)$ is equal to 1. You may find the area by integration, as shown below.

$$\int_{-\infty}^{\infty} f(x)\,dx = \int_0^3 kx\,dx + \int_3^6 k(6 - x)\,dx = 1$$
$$\left[\frac{kx^2}{2}\right]_0^3 + \left[6kx - \frac{kx^2}{2}\right]_3^6 = 1$$

Therefore $\quad \frac{9k}{2} + (36k - 18k) - \left(18k - \frac{9k}{2}\right) = 1$

$$9k = 1$$

So $\quad k = \frac{1}{9}$

Note

In this case you could have found k without integration because the graph of the p.d.f. is a triangle, with area given by $\frac{1}{2} \times$ base $\times$ height, resulting in the equation

$$\frac{1}{2} \times 6 \times k(6 - 3) = 1$$

hence $\quad 9k = 1$

and $\quad k = \frac{1}{9}$.

(b) Sketch the graph of f(x).

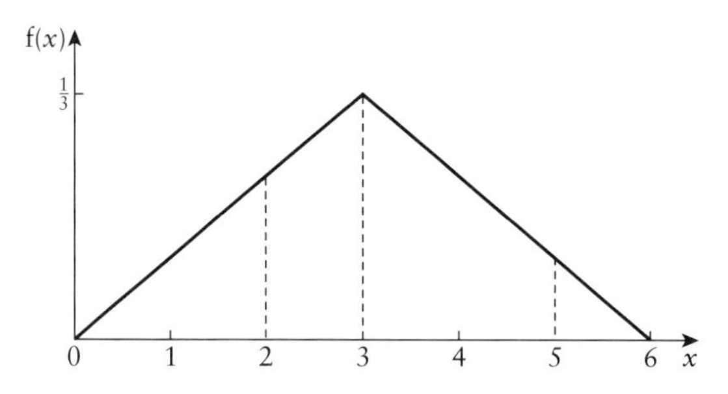

FIGURE 3.5

(c) To find $P(2 \leqslant X \leqslant 5)$, you need to find both $P(2 \leqslant X < 3)$ and $P(3 \leqslant X \leqslant 5)$ because there is a different expression for each part.

$$\begin{aligned}P(2 \leqslant X \leqslant 5) &= P(2 \leqslant X < 3) + P(3 \leqslant X \leqslant 5)\\ &= \int_2^3 \frac{1}{9}x\,dx + \int_3^5 \frac{1}{9}(6-x)\,dx\\ &= \left[\frac{x^2}{18}\right]_2^3 + \left[\frac{2x}{3} - \frac{x^2}{18}\right]_3^5\\ &= \frac{9}{18} - \frac{4}{18} + \left(\frac{10}{3} - \frac{25}{18}\right) - \left(2 - \frac{1}{2}\right)\\ &= 0.72 \text{ to two decimal places.}\end{aligned}$$

The probability that Darren works between 2 and 5 hours in his garden on a randomly selected day is 0.72.

EXERCISE 3A

1 The continuous random variable X has probability density function f(x) where

$$\begin{aligned}f(x) &= kx \quad &&\text{for } 1 \leqslant x \leqslant 6\\ &= 0 &&\text{otherwise.}\end{aligned}$$

(a) Find the value of the constant k.
(b) Sketch $y = f(x)$.
(c) Find $P(X > 5)$.
(d) Find $P(2 \leqslant X \leqslant 3)$.

2 The continuous random variable X has p.d.f. f(x) where

$$\begin{aligned}f(x) &= k(5-x) \quad &&\text{for } 0 \leqslant x \leqslant 4\\ &= 0 &&\text{otherwise.}\end{aligned}$$

(a) Find the value of the constant k.
(b) Sketch $y = f(x)$.
(c) Find $P(1.5 \leqslant X \leqslant 2.3)$.

3 The continuous random variable X has p.d.f. $f(x)$ where

$$f(x) = ax^3 \quad \text{for } 0 \leqslant x \leqslant 3$$
$$= 0 \quad \text{otherwise.}$$

(a) Find the value of the constant a.

(b) Sketch $y = f(x)$.

(c) Find $P(X \leqslant 2)$.

4 The continuous random variable X has p.d.f. $f(x)$ where

$$f(x) = kx \quad \text{for } 0 \leqslant x \leqslant 2$$
$$= 4k - kx \quad \text{for } 2 < x \leqslant 4$$
$$= 0 \quad \text{otherwise.}$$

(a) Find the value of the constant k.

(b) Sketch $y = f(x)$.

(c) Find $P(1 \leqslant X \leqslant 3.5)$.

5 The continuous random variable X has p.d.f. $f(x)$ where

$$f(x) = c \quad \text{for } -3 \leqslant x \leqslant 5$$
$$= 0 \quad \text{otherwise.}$$

(a) Find c.

(b) Sketch $y = f(x)$.

(c) Find $P(|X| < 1)$.

(d) Find $P(|X| > 2.5)$.

6 A continuous random variable X has p.d.f.

$$f(x) = k(x-1)(6-x) \quad \text{for } 1 \leqslant x \leqslant 6$$
$$= 0 \quad \text{otherwise.}$$

(a) Find the value of k.

(b) Sketch $y = f(x)$.

(c) Find $P(2 \leqslant X \leqslant 3)$.

7 A random variable X has p.d.f.

$$f(x) = \begin{cases} (x-1)(2-x) & \text{for } 1 \leqslant x < 2 \\ a & \text{for } 2 \leqslant x \leqslant 4 \\ 0 & \text{otherwise.} \end{cases}$$

(a) Find the value of the constant a.

(b) Sketch $y = f(x)$.

(c) Find $P(1.5 \leqslant X \leqslant 2.5)$.

(d) Find $P(|X - 2| < 1)$.

8 A random variable X has p.d.f.

$$f(x) = \begin{cases} kx(3-x) & \text{for } 0 \leqslant x \leqslant 3 \\ 0 & \text{otherwise.} \end{cases}$$

(a) Find the value of k.

(b) The lifetime (in years) of an electronic component is modelled by this distribution. Two such components are fitted in a radio which will only function if both devices are working. Find the probability that the radio will still function after two years, assuming that their failures are independent.

9 The planning officer in a council needs information about how long cars stay in the car park, and asks the attendant to do a check on the times of arrival and departure of 100 cars. The attendant provides the following data:

Length of stay	Under 1 hour	1–2 hours	2–4 hours	4–10 hours	More than 10 hours
Number of cars	20	14	32	34	0

The planning officer suggests that the length of stay in hours may be modelled by the continuous random variable X with probability density function of the form

$$f(x) = \begin{cases} k(20-2x) & \text{for } 0 \leqslant x \leqslant 10 \\ 0 & \text{otherwise.} \end{cases}$$

(a) Find the value of k.

(b) Sketch the graph of $f(x)$.

(c) According to this model, how many of the 100 cars would be expected to fall into each of the four categories?

(d) Do you think the model fits the data well?

(e) Are there any obvious weaknesses in the model? If you were the planning officer, would you be prepared to accept the model as it is, or would you want any further information?

10 A fish farmer has a very large number of trout in a lake. Before deciding whether to net the lake and sell the fish, she collects a sample of 100 fish and weighs them. The results (in kg) are as follows.

Weight, W	Frequency
$0 < W \leqslant 0.5$	2
$0.5 < W \leqslant 1.0$	10
$1.0 < W \leqslant 1.5$	23
$1.5 < W \leqslant 2.0$	26

Weight, W	Frequency
$2.0 < W \leqslant 2.5$	27
$2.5 < W \leqslant 3.0$	12
$3.0 < W$	0

(a) Illustrate these data on a histogram, with the number of fish on the vertical scale and W on the horizontal scale. Is the distribution of the data symmetrical, positively skewed or negatively skewed?

A friend of the farmer suggests that W can be modelled as a continuous random variable and proposes four possible probability density functions.

$$f_1(w) = \tfrac{2}{9}w(3-w) \qquad f_2(w) = \tfrac{10}{81}w^2(3-w)^2$$

$$f_3(w) = \tfrac{4}{27}w^2(3-w) \qquad f_4(w) = \tfrac{4}{27}w(3-w)^2$$

in each case for $0 \leqslant w \leqslant 3$.

(b) Using your calculator (or otherwise), sketch the curves of the four p.d.f.s and state which one matches the data most closely in general shape.

(c) Use this p.d.f. to calculate the number of fish which that model predicts should fall within each group.

(d) Do you think it is a good model?

11 During a war the crew of an aeroplane has to destroy an enemy railway line by dropping bombs. The distance between the railway line and where the bomb hits the ground is X m, where X has the following p.d.f.

$$f(x) = \begin{cases} 10^{-4}(a+x) & \text{for } -a \leqslant x \leqslant 0 \\ 10^{-4}(a-x) & \text{for } 0 \leqslant x \leqslant a \\ 0 & \text{otherwise.} \end{cases}$$

(a) Find the value of a.

(b) Find $P(50 \leqslant X \leqslant 60)$.

(c) Find $P(|X| < 20)$.

[MEI]

12 A random variable X has a probability density function f given by

$$f(x) = \begin{cases} cx(5-x) & 0 \leqslant x \leqslant 5 \\ 0 & \text{otherwise.} \end{cases}$$

Show that $c = \frac{6}{125}$.

The lifetime X (in years) of an electric light bulb has this distribution. Given that a standard lamp is fitted with two such new bulbs and that their failures are independent, find the probability that neither bulb fails in the first year and the probability that exactly one bulb fails within two years.

[MEI]

13 This graph shows the probability distribution function, $f(x)$, for the heights, X, of waves at the point with Latitude 44°N Longitude 41°W.

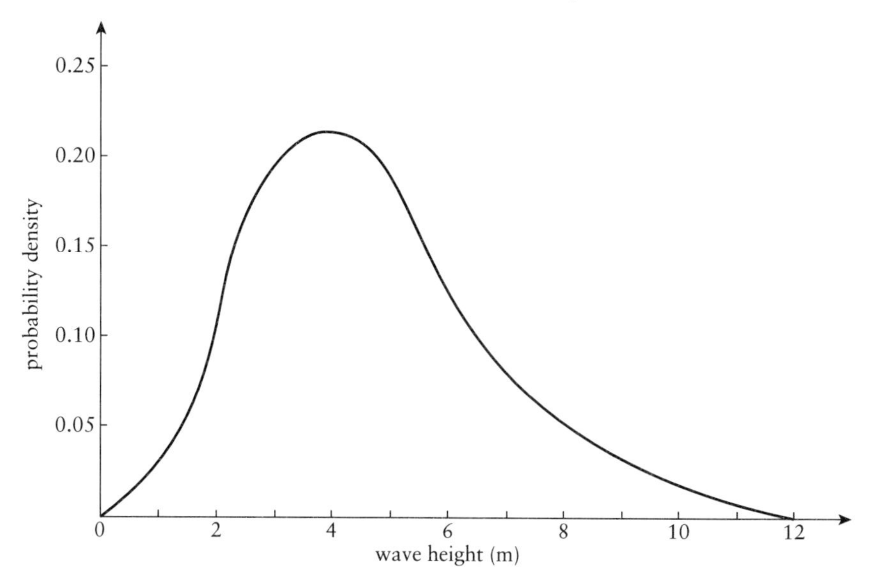

(a) Write down the values of $f(x)$ when $x = 0, 2, 4, \ldots, 12$.

(b) Hence estimate the probability that the height of a randomly selected wave is in the interval

(i) 0–2 m **(ii)** 2–4 m **(iii)** 4–6 m
(iv) 6–8 m **(v)** 8–10 m **(vi)** 10–12 m.

A model is proposed in which

$$f(x) = kx(12 - x)^2 \quad \text{for } 0 \leqslant x \leqslant 12$$
$$= 0 \quad \text{otherwise.}$$

(c) Find the value of k.

(d) Find, according to this model, the probability that a randomly selected wave is in the interval

(i) 0–2 m **(ii)** 2–4 m **(iii)** 4–6 m
(iv) 6–8 m **(v)** 8–10 m **(vi)** 10–12 m.

(e) By comparing the figures from the model with the real data, state whether you think it is a good model or not.

EXPECTATION AND VARIANCE

You will recall that, for a discrete random variable, expectation and variance are given by:

$$\mathrm{E}(X) = \sum_i x_i p_i$$

$$\mathrm{Var}(X) = \sum_i (x_i - \mu)^2 p_i = \sum_i x_i^2 p_i - [\mathrm{E}(X)]^2$$

where μ is the mean and p_i is the probability of the outcome x_i for $i = 1, 2, 3, \ldots$, with the various outcomes covering all possibilities.

The expressions for the expectation and variance of a continuous random variable are equivalent, but with summation replaced by integration.

$$\mathrm{E}(X) = \int_{\substack{\text{All}\\\text{values}\\\text{of } x}} x\,\mathrm{f}(x)\,\mathrm{d}x$$

$$\mathrm{Var}(X) = \int_{\substack{\text{All}\\\text{values}\\\text{of } x}} (x - \mu)^2 \mathrm{f}(x)\,\mathrm{d}x = \int_{\substack{\text{All}\\\text{values}\\\text{of } x}} x^2 \mathrm{f}(x)\,\mathrm{d}x - [\mathrm{E}(X)]^2$$

$\mathrm{E}(X)$ is the same as the population mean, μ, and is often called the mean of X.

EXAMPLE 3.4

The response time, in seconds, for a contestant in a general knowledge quiz is modelled by a continuous random variable X whose p.d.f. is

$$\mathrm{f}(x) = \frac{x}{50} \quad \text{for } 0 < x \leqslant 10.$$

The rules state that a contestant who makes no answer is disqualified from the whole competition. This has the consequence that everybody gives an answer, if only a guess, to every question. Find

(a) the mean time in seconds for a contestant to respond to a particular question

(b) the standard deviation of the time taken.

The organiser estimates the proportion of contestants who are guessing by assuming that they are those whose time is at least one standard deviation greater than the mean.

(c) Using this assumption, estimate the probability that a randomly selected response is a guess.

Solution **(a)** Mean time: $$\mathrm{E}(X) = \int_0^{10} x\frac{x}{50}\,\mathrm{d}x$$

$$= \left[\frac{x^3}{150}\right]_0^{10} = \frac{1000}{150} = \frac{20}{3}$$

$$= 6\tfrac{2}{3}$$

The mean time is $6\frac{2}{3}$ seconds.

(b) Variance: $$\mathrm{Var}(X) = \int_0^{10} x^2\mathrm{f}(x)\,\mathrm{d}x - [\mathrm{E}(X)]^2$$

$$= \int_0^{10} \frac{x^3}{50}\,\mathrm{d}x - (6\tfrac{2}{3})^2$$

$$= \left[\frac{x^4}{200}\right]_0^{10} - (6\tfrac{2}{3})^2$$

$$= 5\tfrac{5}{9}$$

Standard deviation = $\sqrt{\text{Variance}} = \sqrt{5.\dot{5}}$

The standard deviation of the times is 2.357 seconds (to 3 dp).

(c) All those with response times greater than 6.667 + 2.357 = 9.024 seconds are taken to be guessing. The longest possible time is 10 seconds.

The probability that a randomly selected response is a guess is given by

$$\int_{9.024}^{10} \frac{x}{50}\,\mathrm{d}x$$

$$= \left[\frac{x^2}{100}\right]_{9.024}^{10}$$

$$= 0.186$$

So just under 1 in 5 answers are deemed to be guesses.

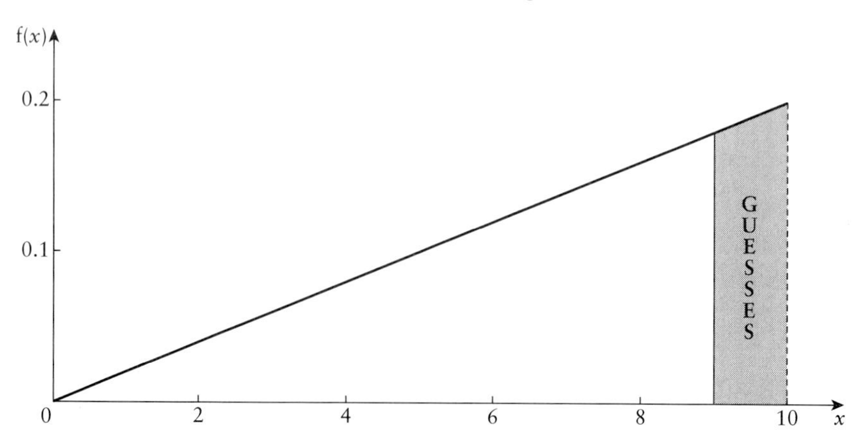

FIGURE 3.6

Note Although the intermediate answers have been given rounded to three decimal places, more figures have been carried forward into subsequent calculations.

EXAMPLE 3.5

The number of hours per day that Darren spends in his garden is modelled (as on page 43) by the continuous random variable X, the p.d.f. of which is

$$\begin{aligned} f(x) &= \frac{1}{9}x && \text{for } 0 \leqslant x \leqslant 3 \\ &= \frac{(6-x)}{9} && \text{for } 3 < x \leqslant 6 \\ &= 0 && \text{otherwise.} \end{aligned}$$

Find $E(X)$, the mean number of hours per day that Darren spends in his garden.

Solution

$$\begin{aligned} E(X) &= \int_{-\infty}^{\infty} x f(x)\,dx \\ &= \int_0^3 x\frac{1}{9}x\,dx + \int_3^6 x\frac{(6-x)}{9}\,dx \\ &= \left[\frac{x^3}{27}\right]_0^3 + \left[\frac{x^2}{3} - \frac{x^3}{27}\right]_3^6 \\ &= 1 + (12-8) - (3-1) \\ &= 3 \end{aligned}$$

Darren spends a mean of 3 hours per day in his garden.

Notice that in this case $E(X)$ can be found from the line of symmetry of the graph of $f(x)$. This situation often arises and you should be alert to the possibility of finding $E(X)$ by symmetry; see figure 3.7.

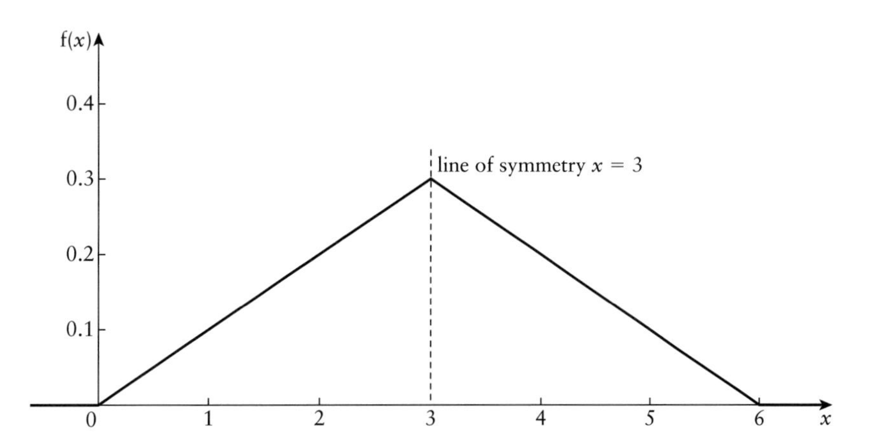

FIGURE 3.7

MEDIAN AND MODE

THE MEDIAN

The median value of a continuous random variable X with p.d.f. $f(x)$ is the value m for which

$$P(X < m) = P(X > m) = 0.5.$$

Consequently $\int_{-\infty}^{m} f(x)\,dx = 0.5$ and $\int_{m}^{\infty} f(x)\,dx = 0.5.$

The median is the value m such that the line $x = m$ divides the area under the curve $f(x)$ into two equal parts. In figure 3.8 a is the smallest possible value of X, b the largest. The line $x = m$ divides the shaded region into two regions A and B, both with area 0.5.

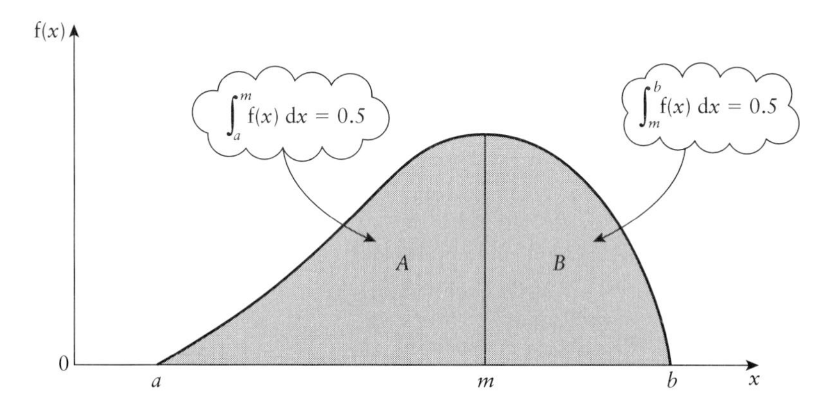

FIGURE 3.8

Note In general the mean does not divide the area into two equal parts but it will do so if the curve is symmetrical about it because, in that case, it is equal to the median.

THE MODE

The mode of a continuous random variable X whose p.d.f. is $f(x)$ is the value for which $f(x)$ has the greatest value. Thus the mode is the value of X where the curve is at its highest.

If the mode is at a local maximum of $f(x)$, then it may often be found by differentiating $f(x)$ and solving the equation

$$f'(x) = 0.$$

EXAMPLE 3.6

The continuous random variable X has p.d.f. $f(x)$ where

$$\begin{aligned} f(x) &= 4x(1 - x^2) \quad \text{for } 0 \leqslant x \leqslant 1 \\ &= 0 \quad \text{otherwise.} \end{aligned}$$

Find **(a)** the mode
(b) the median.

Solution **(a)** The mode is found by differentiating $f(x) = 4x - 4x^3$

$$f'(x) = 4 - 12x^2$$

Solving $f'(x) = 0$ $\qquad x = \frac{1}{\sqrt{3}} = 0.577$ to 3 decimal places.

> $x = -0.577$ is also a root of $f'(x) = 0$ but is outside the range $0 \leqslant x \leqslant 1$

It is easy to see from the shape of the graph (see figure 3.9) that this must be a maximum, and so the mode is 0.577.

(b) The median, m, is given by $\int_{-\infty}^{m} f(x)\,dx = 0$

$$\Rightarrow \int_0^m (4x - 4x^3)\,dx = 0.5$$

> Since $x \geqslant 0$

$$\left[2x^2 - x^4\right]_0^m = 0.5$$

$$2m^2 - m^4 = 0.5$$

Rearranging gives

$$2m^4 - 4m^2 + 1 = 0.$$

This is a quadratic equation in m^2. The formula gives

$$m^2 = \frac{4 \pm \sqrt{16 - 8}}{4}$$

$m = 0.541$ or 1.307 to 3 decimal places.

Since 1.307 is outside the domain of X, the median is 0.541.

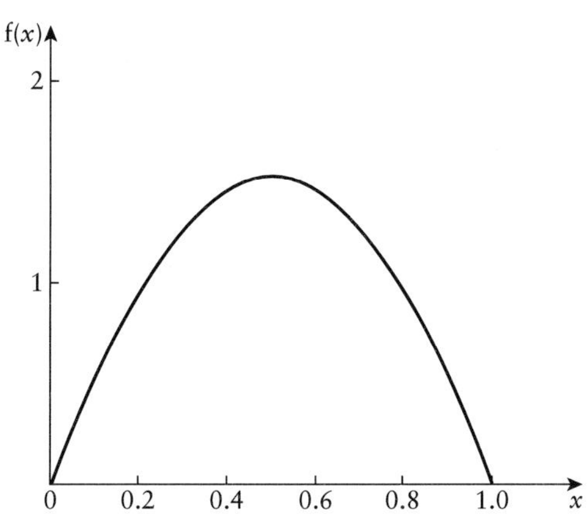

FIGURE 3.9

THE RECTANGULAR DISTRIBUTION

It is common to describe distributions by the shapes of the graphs of their p.d.f.s: U-shaped, J-shaped, etc.

The *rectangular distribution* is particularly simple since its p.d.f. is constant over a range of values and zero elsewhere.

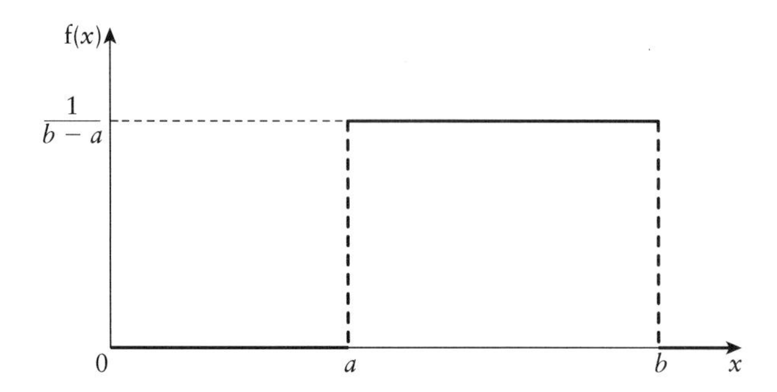

FIGURE 3.10

In figure 3.10, X may take values between a and b, and is zero elsewhere. Since the area under the graph must be 1, the height is $\dfrac{1}{b-a}$. The rectangular distribution is sometimes referred to as the uniform distribution, but this term is more often used when all the outcomes for a *discrete* variable are equally likely, like the score when a single die is thrown.

You will study the rectangular distribution further in Chapter 4.

EXAMPLE 3.7

A junior gymnastics league is open to children who are at least five years old but have not yet had their ninth birthday. The age, X years, of a member is modelled by the rectangular distribution over the range of possible values between five and nine. Age is measured in years and decimal parts of a year, rather than just completed years. Find

(a) the p.d.f. $f(x)$ of X

(b) $P(6 \leqslant X \leqslant 7)$

(c) $E(X)$

(d) $Var(X)$

(e) the percentage of the children whose ages are within one standard deviation of the mean.

Solution **(a)** The p.d.f. $f(x) = \frac{1}{9-5} = \frac{1}{4}$ for $5 \leqslant x < 9$

$= 0$ otherwise.

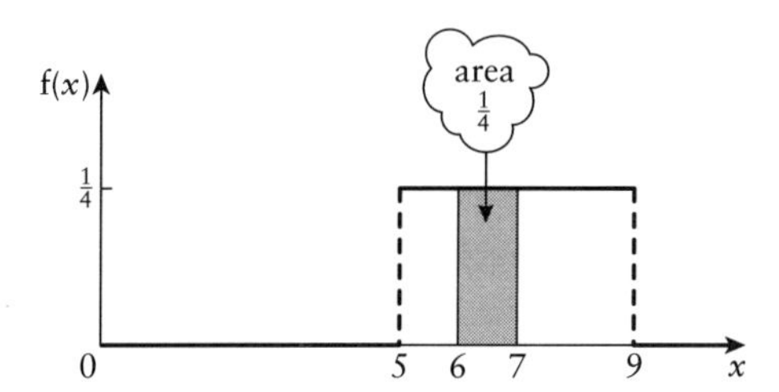

FIGURE 3.11

(b) $P(6 \leqslant X \leqslant 7) = \frac{1}{4}$ by inspection of the rectangle above. Alternatively, using integration

$$P(6 \leqslant X \leqslant 7) = \int_6^7 f(x)\,dx = \int_6^7 \frac{1}{4}\,dx$$

$$= \left[\frac{x}{4}\right]_6^7$$

$$= \frac{7}{4} - \frac{6}{4}$$

$$= \frac{1}{4}.$$

(c) By the symmetry of the graph $E(X) = 7$. Alternatively, using integration

$$E(X) = \int_{-\infty}^{\infty} x\,f(x)\,dx = \int_5^9 \frac{x}{4}\,dx$$

$$= \left[\frac{x^2}{8}\right]_5^9$$

$$= \frac{81}{8} - \frac{25}{8} = 7.$$

(d) $$\text{Var}(X) = \int_{-\infty}^{\infty} x^2 f(x)\,dx - [E(X)]^2 = \int_5^9 \frac{x^2}{4}\,dx - 7^2$$

$$= \left[\frac{x^3}{12}\right]_5^9 - 49$$

$$= \frac{729}{12} - \frac{125}{12} - 49$$

$= 1.333$ to 3 decimal places.

(e) Standard deviation $= \sqrt{\text{Variance}} = \sqrt{1.333} = 1.155.$

So the percentage within 1 standard deviation of the mean is

$$\frac{2 \times 1.155}{4} \times 100\% = 57.7\%.$$

Note For a normal distribution 68% would be within one standard deviation of the mean.

Exercise 3B

1 The continuous random variable X has p.d.f. $f(x)$ where

$$\begin{aligned} f(x) &= \tfrac{1}{8}x \quad \text{for } 0 \leqslant x \leqslant 4 \\ &= 0 \quad \text{otherwise.} \end{aligned}$$

Find

(a) $E(X)$

(b) $\text{Var}(X)$

(c) the median value of X.

2 The continuous random variable T has p.d.f. defined by

$$\begin{aligned} f(t) &= \frac{6-t}{18} \quad \text{for } 0 \leqslant t \leqslant 6 \\ &= 0 \quad \text{otherwise.} \end{aligned}$$

Find

(a) $E(T)$

(b) $\text{Var}(T)$

(c) the median value of T.

3 The continuous random variable Y has p.d.f. $f(y)$ defined by

$$\begin{aligned} f(y) &= 12y^2(1-y) \quad \text{for } 0 \leqslant y \leqslant 1 \\ &= 0 \quad \text{otherwise.} \end{aligned}$$

Find

(a) $E(Y)$

(b) $\text{Var}(Y)$

(c) the median value of Y.

4 The random variable X has p.d.f.

$$\begin{aligned} f(x) &= \tfrac{1}{6} \quad \text{for } -2 \leqslant x \leqslant 4 \\ &= 0 \quad \text{otherwise.} \end{aligned}$$

(a) Sketch the graph of $f(x)$.

(b) Find $P(X < 2)$.

(c) Find $E(X)$.

(d) Find $P(|X| < 1)$.

5 The continuous random variable X has p.d.f. $f(x)$ defined by

$$f(x) = \begin{cases} \frac{2}{9}x(3-x) & \text{for } 0 \leqslant x \leqslant 3 \\ 0 & \text{otherwise.} \end{cases}$$

(a) Find $E(X)$.

(b) Find $Var(X)$.

(c) Find the mode of X.

(d) Find the median value of X.

(e) Draw a sketch graph of $f(x)$ and comment on your answers to parts (a), (c) and (d) in the light of what it shows you.

6 The function $f(x) = \begin{cases} k(3+x) & \text{for } 0 \leqslant x \leqslant 2 \\ 0 & \text{otherwise} \end{cases}$

is the probability density function of the random variable X.

(a) Show that $k = \frac{1}{8}$.

(b) Find the mean and variance of X.

(c) Find the probability that a randomly selected value of X lies between 1 and 2.

7 The distribution of the lengths of adult Martian lizards is uniform between 10 cm and 20 cm. There are no adult lizards outside this range.

(a) Write down the p.d.f. of the lengths of the lizards.

(b) Find the mean and variance of the lengths of the lizards.

(c) What proportion of the lizards have lengths within

(i) one standard deviation of the mean

(ii) two standard deviations of the mean?

8 The p.d.f. of the lifetime, X hours, of a brand of electric light bulb is modelled by

$$f(x) = \begin{cases} \frac{1}{60\,000}x & \text{for } 0 \leqslant x \leqslant 300 \\ \frac{1}{50} - \frac{1}{20\,000}x & \text{for } 300 < x \leqslant 400 \\ 0 & \text{for } x > 400. \end{cases}$$

(a) Sketch the graph of $f(x)$.

(b) Show that $f(x)$ fulfils the conditions for it to be a p.d.f.

(c) Find the expected lifetime of a bulb.

(d) Find the variance of the lifetimes of the bulbs.

(e) Find the probability that a randomly selected bulb will last less than 100 hours.

9 The marks of candidates in an examination are modelled by the continuous random variable X with p.d.f.

$$f(x) = kx(x-50)^2(100-x) \quad \text{for } 0 \leqslant x \leqslant 100$$
$$= 0 \quad \text{otherwise.}$$

(a) Find the value of k.

(b) Sketch the graph of $f(x)$.

(c) Describe the shape of the graph and give an explanation of how such a graph might occur, in terms of the examination and the candidates.

(d) Is it permissible to model a mark, which is a discrete variable going up in steps of 1, by a continuous random variable like X, as defined in this question?

10 The municipal tourism officer at a Mediterranean resort on the Costa Del Sol wishes to model the amount of sunshine per day during the holiday season. She denotes by X the number of hours of sunshine per day between 8 am and 8 pm and she suggests the following probability density function for X:

$$f(x) = k[(x-3)^2 + 4] \quad \text{for } 0 \leqslant x \leqslant 12$$
$$= 0 \quad \text{otherwise.}$$

(a) Show that $k = \frac{1}{300}$ and sketch the graph of the p.d.f. $f(x)$.

(b) Assuming that the model is accurate, find the mean and standard deviation of the number of hours of sunshine per day. Find also the probability of there being more than eight hours of sunshine on a randomly chosen day.

(c) Obtain a cubic equation for m, the median number of hours of sunshine, and verify that m is about 9.74 to 2 decimal places.

[MEI]

11 An examination is taken by a large number of candidates. The marks scored are modelled by the continuous random variable X with probability density function

$$f(x) = kx^3(120-x), \quad 0 \leqslant x \leqslant 100.$$

(You should assume throughout this question that marks are on a continuous scale. Hence there is no need to consider continuity corrections.)

(a) Sketch the graph of this probability density function. What does the model suggest about the abilities of the candidates in relation to this examination?

(b) Show that $k = 10^{-9}$.

(c) The pass mark is set at 50. Find what proportion of candidates fail the examination.

(d) The top 20% of candidates are awarded a distinction. Determine whether a mark of 90 is sufficient for a distinction. Find the least whole number mark which is sufficient for a distinction.

[MEI]

CUMULATIVE DISTRIBUTION FUNCTION

A statistician has been challenged to construct a mathematical model for the times taken by 500 runners to complete a local half-marathon race. The winner completed the race in 1 hour 4 minutes, while everyone managed to complete the course in under 4 hours.

The statistician was given this summary data from which to work:

Time (hours)	Finished (%)
1	0
$1\frac{1}{4}$	3
$1\frac{1}{2}$	15
$1\frac{3}{4}$	33
2	49
$2\frac{1}{4}$	57
$2\frac{1}{2}$	75
3	91
$3\frac{1}{2}$	99
4	100

The statistician proposed a model in which a runner's time, X hours, is a continuous random variable with p.d.f.

$$\begin{aligned} f(x) &= \frac{4}{27}(x-1)(4-x)^2 & 1 \leqslant x \leqslant 4 \\ &= 0 & \text{otherwise} \end{aligned}$$

According to this model the mode is at 2 hours, and everybody finishes in between 1 hour and 4 hours; see figure 3.12.

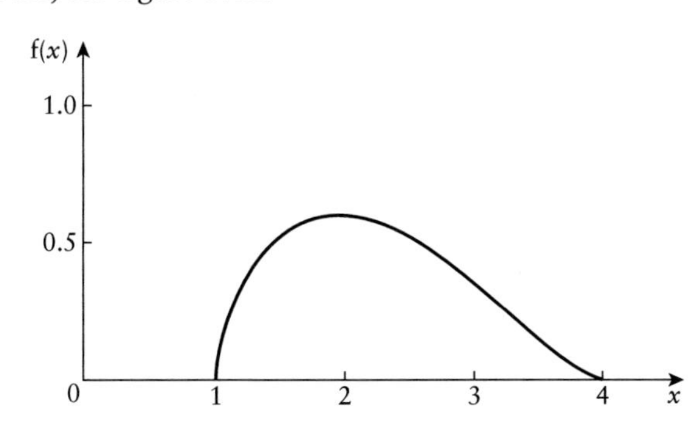

FIGURE 3.12

How does this model compare with the figures you were given for the actual race?

Those figures gave the *cumulative distribution*, the total numbers (expressed as percentages) of runners who had finished by certain times. To obtain the equivalent figures from the model, you must find the relevant area under the graph in figure 3.13.

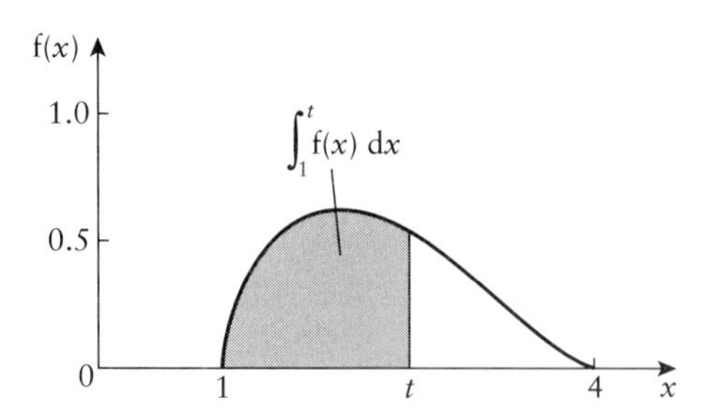

FIGURE 3.13

In this model, the proportion finishing by time t hours is given by

$$\begin{aligned}\int_1^t f(x)\,dx &= \int_1^t \tfrac{4}{27}(x-1)(4-x)^2\,dx \\ &= \tfrac{4}{27}\int_1^t (x^3 - 9x^2 + 24x - 16)\,dx \\ &= \tfrac{4}{27}\left[\tfrac{1}{4}x^4 - 3x^3 + 12x^2 - 16x\right]_1^t \\ &= \tfrac{4}{27}\left(\tfrac{1}{4}t^4 - 3t^3 + 12t^2 - 16t\right) - (-1) \\ &= \tfrac{1}{27}t^4 - \tfrac{4}{9}t^3 + \tfrac{16}{9}t^2 - \tfrac{64}{27}t + 1\end{aligned}$$

This is called the *cumulative distribution function* and denoted by $F(t)$. In this case,

$$\begin{aligned}F(t) &= 0 && \text{for } t < 1 \\ &= \tfrac{1}{27}t^4 - \tfrac{4}{9}t^3 + \tfrac{16}{9}t^2 - \tfrac{64}{27}t + 1 && \text{for } 1 \leqslant t \leqslant 4 \\ &= 1 && \text{for } t > 4.\end{aligned}$$

To find the proportions of runners finishing by any time, substitute that value for t; so when $t = 2$

$$\begin{aligned}F(2) &= \tfrac{1}{27} \times 2^4 - \tfrac{4}{9} \times 2^3 + \tfrac{16}{9} \times 2^2 - \tfrac{64}{27} \times 2 + 1 \\ &= 0.41 \text{ to 2 decimal places.}\end{aligned}$$

Here is the complete table, with all the values worked out.

Time (hours)	Model	Runners
1.00	0.00	0.00
1.25	0.04	0.03
1.50	0.13	0.15
1.75	0.26	0.33
2.00	0.41	0.49
2.25	0.55	0.57
2.50	0.69	0.75
3.00	0.89	0.91
3.50	0.98	0.99
4.00	1.00	1.00

Notice the distinctive shape of the curves of these functions (figure 3.14), sometimes called an *ogive*. You have probably met this already, when drawing cumulative frequency curves, for example in *Statistics 1*.

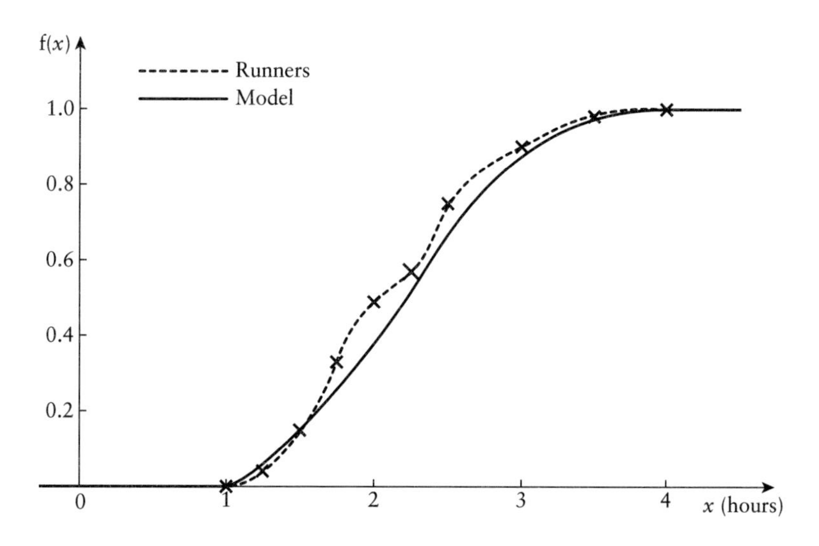

FIGURE 3.14

Notes

1 Notice the use of lower and upper case letters here. The probability density function is denoted by the lower case f, whereas the cumulative distribution function is given the upper case F.

2 F was derived here as a function of t rather than x, to avoid using the same variable in the expression to be integrated. In this case t was a natural variable to use because time was involved, but that is not always the case.

It is more usual to write F as a function of x, $\mathrm{F}(x)$, but you would not be correct to write down an expression like

$$\mathrm{F}(x) = \int_1^x \frac{4}{27}(x-1)(4-x)^2\,\mathrm{d}x \qquad \text{INCORRECT}$$

since x would then be both a limit of the integral and the variable used within it.

To overcome this problem a dummy variable, u, is used in the rest of this section, so that $\mathrm{F}(x)$ is now written,

$$\mathrm{F}(x) = \int_1^x \frac{4}{27}(u-1)(4-u)^2\,\mathrm{d}u \qquad \text{CORRECT}$$

You may of course use another symbol, like y or p, rather than u, anything except x.

3 The term cumulative distribution function is often abbreviated to c.d.f.

Properties of the cumulative distribution function, F(*x*)

The graphs, figure 3.15, show the probability density function $\mathrm{f}(x)$ and the cumulative distribution function $\mathrm{F}(x)$ of a typical continuous random variable X. You will see that the values of the random variable always lie between a and b.

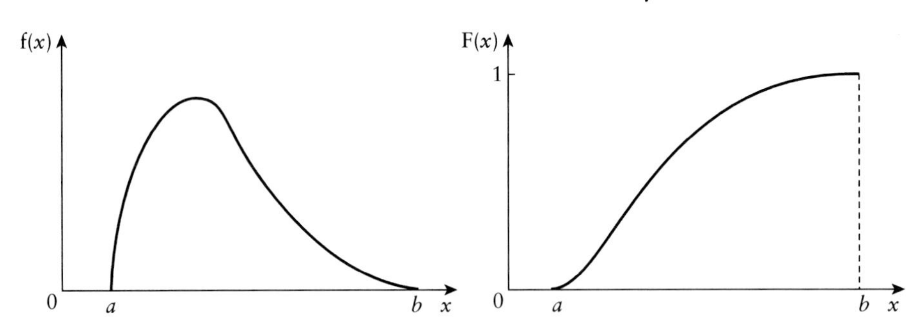

FIGURE 3.15

These graphs illustrate a number of general results for cumulative distribution functions.

1 $\mathrm{F}(x) = 0$ for $x \leqslant a$, the lower limit of x.

The probability of X taking a value less than or equal to a is zero; the value of X must be greater than or equal to a.

2 $\mathrm{F}(x) = 1$ for $x \geqslant b$, the upper limit of x.
X cannot take values greater than b.

3 $P(c \leqslant X \leqslant d) = F(d) - F(c)$
$P(c \leqslant X \leqslant d) = P(X \leqslant d) - P(X \leqslant c)$

This is very useful when finding probabilities from a p.d.f. or a c.d.f.

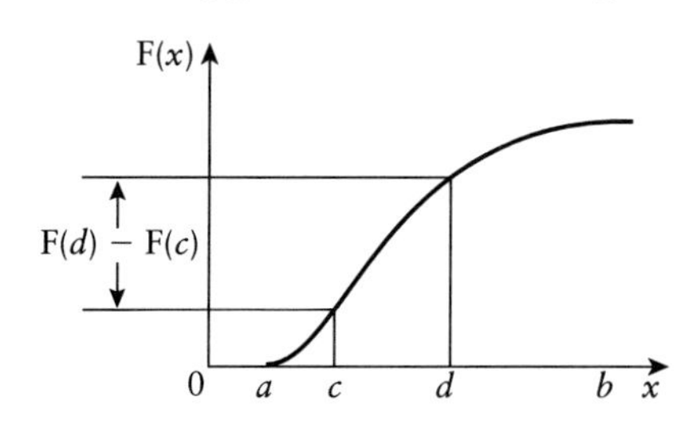

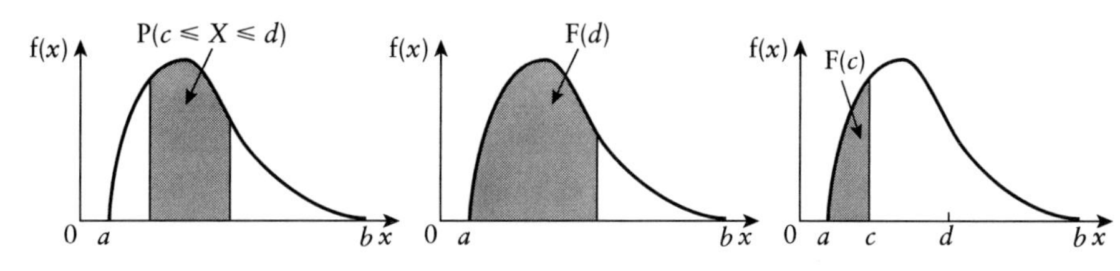

FIGURE 3.16

4 The median, m, satisfies the equation $F(m) = 0.5$.
$P(X \leqslant m) = 0.5$ by definition of the median.

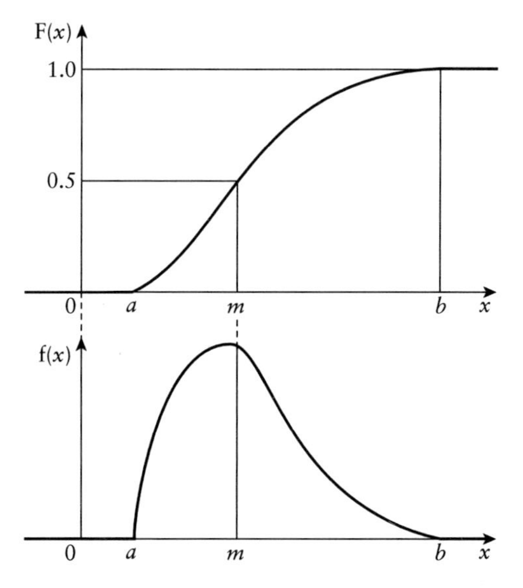

FIGURE 3.17

5 $f(x) = \dfrac{d}{dx}F(x) = F'(x)$

Since you integrate $f(x)$ to obtain $F(x)$, the reverse must also be true: differentiating $F(x)$ gives $f(x)$.

EXAMPLE 3.8

A machine saws planks of wood to a nominal length. The continuous random variable X represents the error in millimetres of the actual length of a plank coming off the machine. The variable X has p.d.f. $f(x)$ where

$$f(x) = \frac{10 - x}{50} \quad \text{for } 0 \leqslant x \leqslant 10$$
$$= 0 \quad \text{otherwise.}$$

(a) Sketch $f(x)$.
(b) Find the cumulative distribution function $F(x)$.
(c) Sketch $F(x)$ for $0 \leqslant x \leqslant 10$.
(d) Find $P(2 \leqslant X \leqslant 7)$.
(e) Find the median value of X.

A customer refuses to accept planks for which the error is greater than 8 mm.
(f) What percentage of planks will he reject?

Solution **(a)**

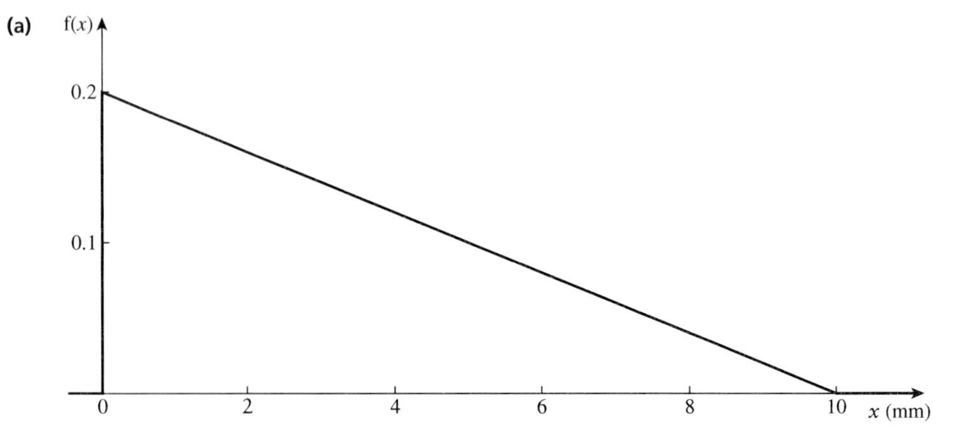

FIGURE 3.18

(b) $F(x) = \displaystyle\int_0^x \frac{(10 - u)}{50}\, du$

$$= \frac{1}{50}\left[10u - \frac{u^2}{2}\right]_0^x$$

$$= \frac{1}{5}x - \frac{1}{100}x^2$$

The full definition of $F(x)$ is:

$$F(x) = 0 \quad \text{for } x < 0$$
$$= \frac{1}{5}x - \frac{1}{100}x^2 \quad \text{for } 0 \leqslant x \leqslant 10$$
$$= 1 \quad \text{for } x > 10.$$

(c) The graph F(x) is shown in figure 3.19.

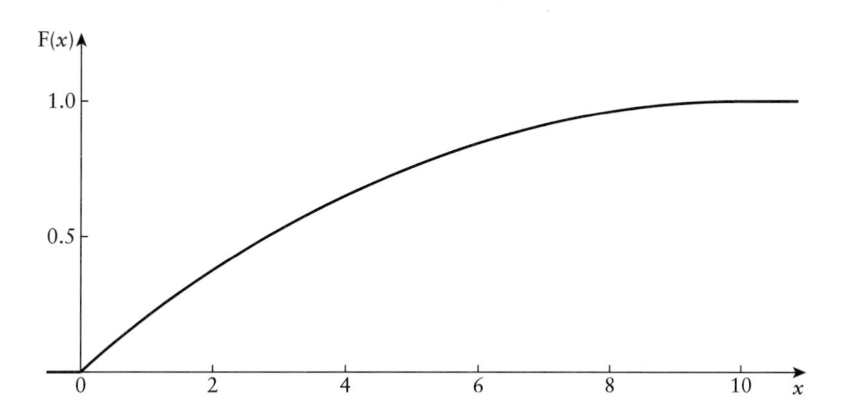

FIGURE 3.19

(d) $P(2 \leqslant X \leqslant 7) = F(7) - F(2)$

$$= \left[\frac{7}{5} - \frac{49}{100}\right] - \left[\frac{2}{5} - \frac{4}{100}\right]$$

$$= 0.91 - 0.36$$

$$= 0.55.$$

(e) The median value of X is found by solving the equation

$$F(m) = 0.5$$

$$\frac{1}{5}m - \frac{1}{100}m^2 = 0.5.$$

This is rearranged to give

$$m^2 - 20m + 50 = 0$$

$$m = \frac{20 \pm \sqrt{20^2 - 4 \times 50}}{2}$$

$m = 2.93$ (or 17.07, outside the domain for X).

The median error is 2.93 mm.

(f) The customer rejects those planks for which $8 \leqslant X \leqslant 10$

$$P(8 \leqslant X \leqslant 10) = F(10) - F(8)$$

$$= 1 - 0.96$$

so 4% of planks are rejected.

EXAMPLE 3.9

The p.d.f. of a continuous random variable X is given by:

$$\begin{aligned} f(x) &= \frac{x}{24} && \text{for } 0 \leqslant x \leqslant 4 \\ &= \frac{(12 - x)}{48} && \text{for } 4 \leqslant x \leqslant 12 \\ &= 0 && \text{otherwise.} \end{aligned}$$

(a) Sketch $f(x)$.
(b) Find the cumulative distribution function $F(x)$.
(c) Sketch $F(x)$.

Solution **(a)** The graph of $f(x)$ is shown in figure 3.20.

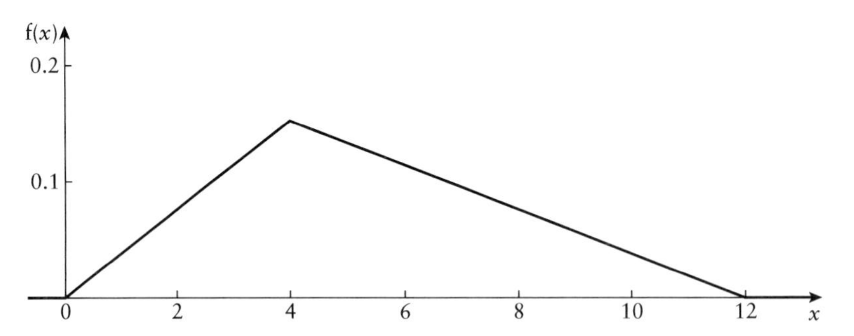

FIGURE 3.20

(b) For $0 \leqslant x \leqslant 4$, $F(x) = \int_0^x \frac{u}{24}\,du$

$$= \left[\frac{u^2}{48}\right]_0^x$$

$$= \frac{x^2}{48}$$

and so $F(4) = \frac{1}{3}$.

For $4 \leqslant x \leqslant 12$, a second integration is required:

$$F(x) = \int_0^4 \frac{u}{24}\,du + \int_4^x \left(\frac{12 - u}{48}\right) du$$

$$= F(4) + \left[\frac{u}{4} - \frac{u^2}{96}\right]_4^x$$

$$= \frac{1}{3} + \frac{x}{4} - \frac{x^2}{96} - \frac{5}{6}$$

$$= -\frac{1}{2} + \frac{x}{4} - \frac{x^2}{96}$$

So the full definition of F(x) is

$$\begin{aligned} F(x) &= 0 && \text{for } x < 0 \\ &= \frac{x^2}{48} && \text{for } 0 \leqslant x \leqslant 4 \\ &= -\frac{1}{2} + \frac{x}{4} - \frac{x^2}{96} && \text{for } 4 \leqslant x \leqslant 12 \\ &= 1 && \text{for } x > 12. \end{aligned}$$

(c) The graph of F(x) is shown in figure 3.21.

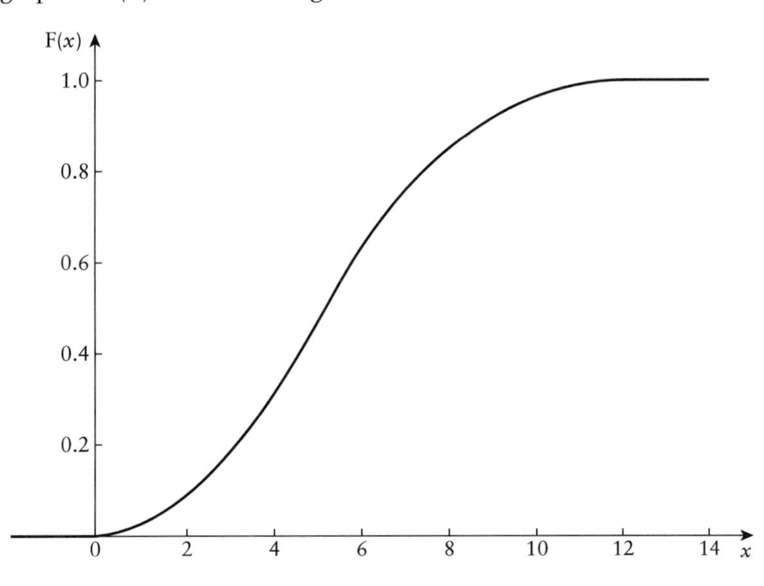

FIGURE 3.21

EXAMPLE 3.10

The continuous random variable X has cumulative distribution function F(x) given by:

$$F(x) = \begin{cases} 0 & \text{for } x < 2 \\ \dfrac{x^2}{32} - \dfrac{1}{8} & \text{for } 2 \leqslant x \leqslant 6 \\ 1 & \text{for } x > 6. \end{cases}$$

Find the p.d.f. f(x).

Solution

$$f(x) = \frac{d}{dx} F(x)$$

$$f(x) = \begin{cases} \dfrac{d}{dx} F(x) = 0 & \text{for } x < 2 \\ \dfrac{d}{dx} F(x) = \dfrac{x}{16} & \text{for } 2 \leqslant x \leqslant 6 \\ \dfrac{d}{dx} F(x) = 0 & \text{for } x > 6. \end{cases}$$

EXERCISE 3C

1 The continuous random variable X has p.d.f. $f(x)$ where

$$f(x) = 0.2 \quad \text{for } 0 \leqslant x \leqslant 5$$
$$= 0 \quad \text{otherwise.}$$

(a) Find $E(X)$.

(b) Find the cumulative distribution function, $F(x)$.

(c) Find $P(0 \leqslant x \leqslant 2)$ using **(i)** $F(x)$ **(ii)** $f(x)$ and show your answer is the same by each method.

2 The continuous random variable U has p.d.f. $f(u)$ where

$$f(u) = ku \quad \text{for } 5 \leqslant u \leqslant 8$$
$$= 0 \quad \text{otherwise.}$$

(a) Find the value of k.

(b) Sketch $f(u)$.

(c) Find $F(u)$.

(d) Sketch the graph of $F(u)$.

3 A continuous random variable X has p.d.f. $f(x)$ where

$$f(x) = cx^2 \quad \text{for } 1 \leqslant x \leqslant 4$$
$$= 0 \quad \text{otherwise.}$$

(a) Find the value of c.

(b) Find $F(x)$.

(c) Find the median of X.

(d) Find the mode of X.

4 The continuous random variable X has p.d.f. $f(x)$ given by

$$f(x) = \frac{k}{(x+1)^4} \quad \text{for } x \geqslant 0$$
$$= 0 \quad \text{for } x < 0$$

where k is a constant.

(a) Show that $k = 3$, and find the cumulative distribution function.

(b) Find the value of x such that $P(X < x) = \frac{7}{8}$.

[Cambridge]

5 The continuous random variable X has c.d.f. given by

$$F(x) = \begin{cases} 0 & \text{for } x < 0 \\ 2x - x^2 & \text{for } 0 \leqslant x \leqslant 1 \\ 1 & \text{for } x > 1. \end{cases}$$

(a) Find $P(X > 0.5)$.

(b) Find the value of q such that $P(X < q) = \frac{1}{4}$.

(c) Find the p.d.f. $f(x)$ of X, and sketch its graph.

[Cambridge]

6 The continuous random variable X has p.d.f. $f(x)$ given by

$$f(x) = \begin{cases} k(4 - x^2) & \text{for } 0 \leqslant x \leqslant 2 \\ 0 & \text{otherwise} \end{cases}$$

where k is a constant.

Show that $k = \frac{3}{16}$ and find the values of $E(X)$ and $Var(X)$.

Find the cumulative distribution function for X, and verify by calculation that the median value of X is between 0.69 and 0.70.

[Cambridge]

7 A random variable X has p.d.f. $f(x)$ where

$$f(x) = 12x^2(1 - x) \quad \text{for } 0 \leqslant x \leqslant 1$$

and $\quad f(x) = 0 \quad$ for all other x.

Find μ, the mean of X, and show that σ, the standard deviation of X, is $\frac{1}{5}$.
Show that $F(x)$, the probability that $X \leqslant x$ (for any value of x between 0 and 1), satisfies

$$F(x) = \begin{cases} 0 & \text{for } x < 0 \\ 4x^3 - 3x^4 & \text{for } 0 \leqslant x \leqslant 1 \\ 1 & \text{for } x > 1. \end{cases}$$

Use this result to show that $P(|X - \mu| < \sigma) = 0.64$.

What would this probability be if, instead, X were normally distributed?

[MEI]

8 The temperature in degrees Celsius in a refrigerator which is operating properly has probability density function given by

$$f(t) = \begin{cases} kt^2(12 - t) & 0 < t < 12 \\ 0 & \text{otherwise.} \end{cases}$$

(a) Show that the value of k is $\frac{1}{1728}$.

(b) Find the cumulative distribution function $F(t)$.

(c) Show, by substitution, that the median temperature is about 7.37 °C.

(d) The temperature in a refrigerator is too high if it is over 10 °C. Find the probability that this occurs.

[MEI]

9 The probability that a randomly chosen flight from Stanston Airport is delayed by more than x hours is

$$\frac{(x - 10)^2}{100} \quad \text{for } 0 \leqslant x \leqslant 10.$$

No flights leave early, and none is delayed for more than 10 hours. The delay, in hours, for a randomly chosen flight is denoted by X.

(a) Find the median, m, of X, correct to three significant figures.

(b) Find the cumulative distribution function, F, of X and sketch the graph of F.

(c) Find the probability density function, f, of X, and sketch the graph of f.

(d) Show that $E(X) = \frac{10}{3}$.

[Cambridge]

10 On any day, the amount of time, measured in hours, that Mr Goggle spends watching television is a continuous random variable T, with cumulative distribution function given by

$$F(t) = \begin{cases} 0 & t < 0 \\ 1 - k(15 - t)^2 & 0 \leqslant t \leqslant 15 \\ 1 & t > 15 \end{cases}$$

where k is a constant.

(a) Show that $k = \frac{1}{225}$ and find $P(5 \leqslant T \leqslant 10)$.

(b) Show that, for $0 \leqslant t \leqslant 15$, the probability density function of T is given by

$$f(t) = \frac{2}{15} - \frac{2t}{225}.$$

(c) Find the median of T.

(d) Find $\text{Var}(T)$.

[Cambridge]

EXERCISE 3D Examination-style questions

1 A computer is used to add up a series of numbers. Each addition introduces an error which may be regarded as a random variable, X, which has the rectangular distribution on the interval $[-a, a]$, where a is, of course, very small. The probability density function for X is illustrated below.

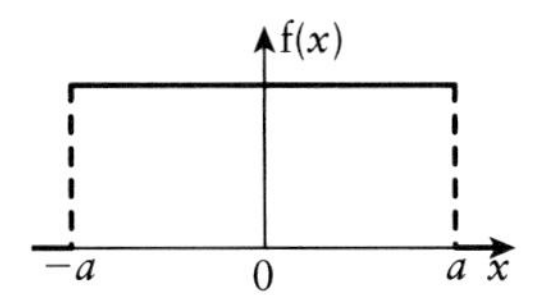

Find, in terms of a,

(a) $f(x)$, the probability density function for X

(b) $E(X)$ and $\text{Var}(X)$.

[MEI, part]

2 The continuous random variable X has probability density function

$$f(x) = \tfrac{3}{1024}x(x-8)^2 \quad 0 \leqslant x \leqslant 8.$$

A sketch of f(x) is shown in the diagram.

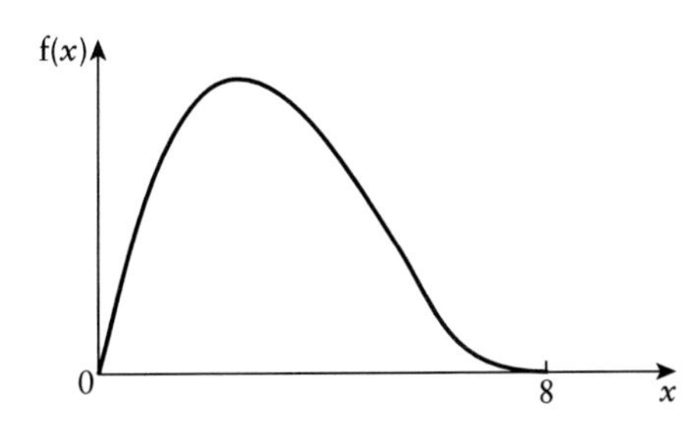

(a) Find E(X) and show that Var(X) = 2.56.

The times, in minutes, taken by a doctor to see her patients are modelled by the continuous random variable $T = X + 2$.

(b) Sketch the distribution of T and describe in words what this model implies about the lengths of the doctor's appointments.

[MEI, part]

3 A continuous random variable X has probability density function f(x). The probability that $X \leqslant x$ is given by the function F(x).

Explain why $F'(x) = f(x)$.

A rod of length $2a$ is broken into two parts at a point whose position is random. State the form of the probability distribution of the length of the smaller part, and state also the mean value of this length.

Two equal rods, each of length $2a$, are broken into two parts at points whose positions are random. X is the length of the shortest of the four parts thus obtained. Find the probability, F(x), that $X \leqslant x$, where $0 < x \leqslant a$.
Hence, or otherwise, show that the probability density function of X is given by

$$f(x) = \frac{2(a-x)}{a^2} \quad \text{for } 0 < x \leqslant a$$
$$= 0 \quad \text{for } x \leqslant 0,\ x > a.$$

Show that the mean value of X is $\frac{1}{3}a$.

Write down the mean value of the sum of the two smaller parts and show that the mean values of the four parts are in the proportions 1 : 2 : 4 : 5.

[JMB]

4 A gardener is attempting to light a bonfire. The time, in minutes, for which a taper will stay alight on a calm day is modelled by the random variable T. The cumulative distribution function of T is given by

$$F(t) = P(T \leqslant t) = \begin{cases} 0 & t < 0 \\ \frac{1}{2}t^3 - \frac{3}{16}t^4 & 0 \leqslant t \leqslant 2 \\ 1 & t > 2. \end{cases}$$

(a) Find $P(T \leqslant 1)$.

(b) Verify that the median m of T satisfies $1.22 < m < 1.23$.

(c) Find the probability density function $f(t)$ of T.

(d) Find the modal time for which a taper will stay alight.

(e) Sketch the probability density function of T.

(f) Give a reason why this model may not be applicable on a windy day and give a sketch of a probability density function that may be more suitable in such conditions.

[Edexcel]

5 Two friends make regular telephone calls to each other. The duration, in minutes, of their telephone conversation is modelled initially by the random variable T, having probability density function

$$F(t) = \begin{cases} \frac{1}{150}(25 - t) & 5 \leqslant t \leqslant 15 \\ 0 & \text{otherwise.} \end{cases}$$

(a) Sketch the probability density function of T.

(b) For all values of t, find the cumulative distribution function of T.

(c) Find the probability that a telephone conversation lasts longer than 12 minutes.

(d) Show that the median duration of a telephone conversation is given by $(25 - 5\sqrt{10})$ minutes.

(e) Give a reason why this initial model may not be realistic for the distribution of durations of telephone conversations.

(f) Sketch the probability density function of a more realistic model.

[Edexcel]

6 Workers on a large industrial estate have journey times to work, in hours, which are modelled by the random variable X with probability density function

$$f(x) = 20x^3(1 - x) \qquad 0 \leqslant x \leqslant 1.$$

(a) Sketch the graph of the probability density function.

(b) Hence state, with an explanation, the likely location of the industrial estate in relation to local housing.

(c) Find the expectation and the standard deviation of the workers' journey times *in minutes*.

(d) Obtain the cumulative distribution function, and use it to verify that the median journey time is a little over 41 minutes.

[MEI]

7 A teacher of young children is thinking of asking her class to guess her height in metres. The teacher considers that the height guessed by a randomly selected child can be modelled by the random variable H with probability density function

$$f(h) = \begin{cases} \frac{3}{16}(4h - h^2) & 0 \leqslant h \leqslant 2 \\ 0 & \text{otherwise.} \end{cases}$$

Using this model,

(a) find $P(H < 1)$

(b) show that $E(H) = 1.25$.

A friend of the teacher suggests that the random variable X with probability density function

$$g(x) = \begin{cases} kx^3 & 0 \leqslant x \leqslant 2 \\ 0 & \text{otherwise,} \end{cases}$$

where k is a constant, might be a more suitable model.

(c) Show that $k = \frac{1}{4}$.

(d) Find $P(X < 1)$.

(e) Find $E(X)$.

(f) Using your calculations in parts **(a)**, **(b)**, **(d)** and **(e)**, state, giving reasons, which of the random variables H or X is likely to be the more appropriate model in this instance.

[Edexcel]

8 The times, in excess of 2 hours, taken to complete a marathon road race are modelled by the continuous random variable T hours, where T has the probability density function

$$f(t) = \begin{cases} \frac{4}{27}t^2(3 - t) & 0 \leqslant t \leqslant 3 \\ 0 & \text{otherwise.} \end{cases}$$

The diagram shows a sketch of the probability density function.

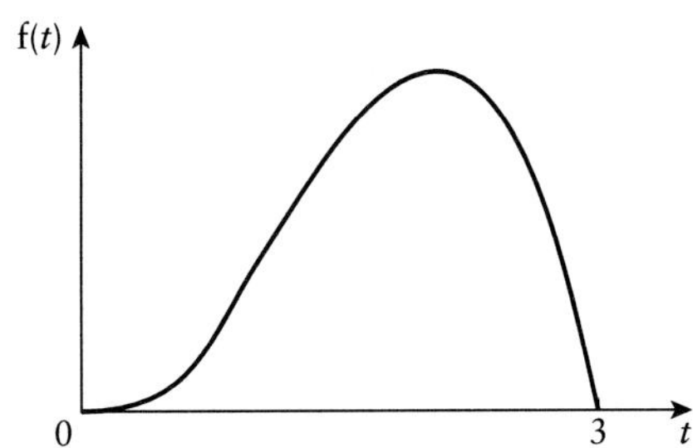

(a) Find the mean and variance of the times taken to complete the race.

(b) Find the modal time taken to complete the race.

(c) What proportion of competitors complete the race in less than the modal time?

(d) Show that the median time to complete the race lies between the mean and the mode.

[MEI]

9 A chemical factory has the capacity to produce up to 1 tonne per day of a particular chemical. The actual amount produced per day varies and is given by the random variable X with probability density function

$$f(x) = \begin{cases} 2x & 0 \leqslant x \leqslant 1 \\ 0 & \text{elsewhere.} \end{cases}$$

(a) Find the mean and variance of X.

(b) Find the cumulative distribution function $F(x) = P(X \leqslant x)$.

All of the chemical that is produced is sold at a profit of £4000 per tonne, but in addition there is a fixed overhead charge of £1000 per day. Thus the daily profit from sales of this chemical, in thousands of pounds, is $Y = 4X - 1$.

(c) State the mean and variance of Y.

(d) Show that the cumulative distribution function for Y is

$P(Y \leqslant y) = \left(\dfrac{y+1}{4}\right)^2$. State the interval of values of y for which this is valid.

(e) Obtain the median profit per day.

[MEI]

10 The continuous random variable X has probability density function $f(x)$ given by

$$f(x) = \begin{cases} \frac{1}{20}x^3 & 1 \leqslant x \leqslant 3 \\ 0 & \text{otherwise.} \end{cases}$$

(a) Sketch $f(x)$ for all values of x.

(b) Calculate $E(X)$.

(c) Show that the standard deviation of X is 0.459 to 3 decimal places.

(d) Show that for $1 \leqslant x \leqslant 3$, $P(X \leqslant x)$ is given by $\frac{1}{80}(x^4 - 1)$ and specify fully the cumulative distribution function of X.

(e) Find the interquartile range for the random variable X.

Some statisticians use the following formula to estimate the interquartile range:

$$\text{interquartile range} = \tfrac{4}{3} \times \text{standard devation.}$$

(f) Use this formula to estimate the interquartile range in this case, and comment.

[Edexcel]

KEY POINTS

1 If X is a continuous random variable with p.d.f. $f(x)$

- $\int f(x)\,dx = 1$
- $f(x) \geqslant 0$ for all x
- $P(c \leqslant X \leqslant d) = \int_c^d f(x)\,dx$
- $E(X) = \int x f(x)\,dx$
- $Var(X) = \int x^2 f(x)\,dx - [E(X)]^2$
- The mode of X is the value for which $f(x)$ has its greatest magnitude.

2 If $g[X]$ is a function of X then

- $E(g[X]) = \int g[x] f(x)\,dx$
- $Var(g[X]) = \int (g[x])^2 f(x)\,dx - [E(g[X])]^2$

3 The cumulative distribution function

- $F(x) = \int_a^x f(u)\,du$ where the constant a is the lower limit of X.
- $f(x) = \dfrac{d}{dx} F(x)$
- For the median, m, $F(m) = 0.5$

4 The rectangular distribution over the interval (a, b)

- $f(x) = \dfrac{1}{b - a}$

Chapter four

CONTINUOUS DISTRIBUTIONS

THE CONTINUOUS UNIFORM DISTRIBUTION

Ray Lowder is an oil-pipe repair man. He has been given the task of locating and repairing a weak spot in a piece of pipeline which is 5 km long. He has been told that the weak spot is just as likely to be found in any part of the pipeline as any other, so he decides to start searching for it at one end, and to continue along the pipeline towards the other end until he finds the weak spot.

Ray can inspect 10 m of pipeline per minute. How long, on average, will it take him to locate such a weak spot? What would be the corresponding standard deviation?

To answer these questions, you need to set up a simple probability model. Let X denote the distance travelled along the pipeline until the weak spot is found.

The probability density function, $f(x)$, will be rectangular, since the weak spot is just as likely to be found at any place along the pipeline. The area within the rectangle must be 1, in order to meet the requirements of a p.d.f., as shown in figure 4.1.

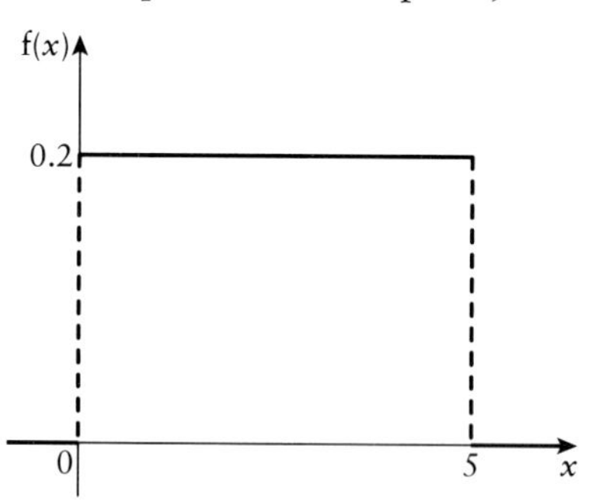

FIGURE 4.1

The rectangle must have a height of 0.2 in order for the area to be 1. Thus the full p.d.f. is given by:

$$f(x) = \begin{cases} 0.2 & 0 \leqslant x \leqslant 5 \\ 0 & \text{otherwise.} \end{cases}$$

To find the 'average' value of X you need to find its mean, or expectation, $\mathrm{E}(X)$.

$$\begin{aligned}\mathrm{E}(X) &= \int x\mathrm{f}(x)\,\mathrm{d}x \\ &= \int_0^5 0.2x\,\mathrm{d}x \\ &= \left[0.1x^2\right]_0^5 \\ &= 2.5 - 0 \\ &= 2.5\end{aligned}$$

This result is, in fact, obvious from the symmetry of the graph of $\mathrm{f}(x)$.

A similar integral gives the variance:

$$\begin{aligned}\mathrm{Var}(X) &= \int x^2\mathrm{f}(x)\,\mathrm{d}x - [\mathrm{E}(X)]^2 \\ &= \int_0^5 0.2x^2\,\mathrm{d}x - 2.5^2 \\ &= \left[\frac{0.2x^3}{3}\right]_0^5 - 6.25 \\ &= 8.\dot{3} - 0 - 6.25 \\ &= 2.08\dot{3}\end{aligned}$$

and so the standard deviation is $\sqrt{\mathrm{Var}(X)} = 1.443$ km (3 d.p.).

Since Ray can inspect 10 m of pipeline per minute, it will take him a mean of 250 minutes to find the weak spot with a standard deviation of 144 minutes.

Ray's oil-pipe problem illustrates a widely used distribution in statistics, namely the **continuous uniform distribution**. This may be summarised as follows:

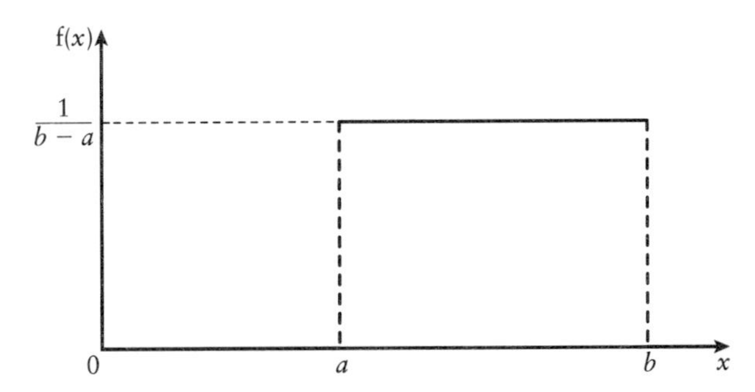

For the uniform continuous distribution over the interval $[a, b]$,

$$\mathrm{f}(x) = \frac{1}{b-a}$$
$$\mathrm{E}(X) = \tfrac{1}{2}(a+b)$$
$$\mathrm{Var}(X) = \tfrac{1}{12}(b-a)^2$$

FIGURE 4.2

These results are given in the Edexcel formula booklet and may be quoted where appropriate in short questions. You do, however, need to know how to derive these results, and that for the cumulative distribution function $\mathrm{F}(x)$, when required.

Derivation of E(X), Var(X) and F(x) for a continuous uniform distribution

E(X)

Let X be uniformly distributed over the interval $[a, b]$.

In order for the area to be 1, the p.d.f. must be:

$$f(x) = \begin{cases} \dfrac{1}{b-a} & a \leqslant x \leqslant b \\ 0 & \text{otherwise.} \end{cases}$$

Then
$$\begin{aligned}
E(X) &= \int x f(x)\,dx \\
&= \int_a^b \frac{x}{b-a}\,dx \\
&= \left[\frac{x^2}{2(b-a)}\right]_a^b \\
&= \frac{b^2}{2(b-a)} - \frac{a^2}{2(b-a)} \\
&= \frac{b^2 - a^2}{2(b-a)} \\
&= \frac{(b+a)(b-a)}{2(b-a)} \\
&= \frac{a+b}{2}
\end{aligned}$$

Var(X)

$$\begin{aligned}
\text{Var}(X) &= \int x^2 f(x)\,dx - [E(X)]^2 \\
&= \int_a^b \frac{x^2}{b-a}\,dx - [E(X)]^2 \\
&= \left[\frac{x^3}{3(b-a)}\right]_a^b - \left[\frac{a+b}{2}\right]^2 \\
&= \frac{b^3 - a^3}{3(b-a)} - \left[\frac{a+b}{2}\right]^2 \\
&= \frac{(b-a)(b^2 + ab + a^2)}{3(b-a)} - \left[\frac{a+b}{2}\right]^2 \\
&= \frac{b^2 + ab + a^2}{3} - \frac{a^2 + 2ab + b^2}{4} \\
&= \frac{4b^2 + 4ab + 4a^2 - 3a^2 - 6ab - 3b^2}{12} \\
&= \frac{b^2 - 2ab + a^2}{12} \\
&= \frac{(b-a)^2}{12}
\end{aligned}$$

F(*X*)

$$F(x) = \int f(x)\,dx$$
$$= \int \frac{1}{b-a}\,dx$$
$$= \frac{x}{b-a} + c$$

Since $F(a) = 0$

$$0 = \frac{a}{b-a} + c \quad \text{and so} \quad c = \frac{-a}{b-a}$$

Thus $F(x) = \dfrac{x-a}{b-a}$

The full c.d.f. is given by:

$$F(x) = \begin{cases} 0 & x < a \\ \dfrac{x-a}{b-a} & a \leqslant x \leqslant b \\ 1 & x > b \end{cases}$$

EXAMPLE 4.1

A meteor watcher sees one meteor (which lasts for a fraction of a second) between 11.00 and 11.15 pm, but forgets to record the precise time at which he sees it. Assuming a uniform distribution over this interval, find:

(a) the mean time at which he is likely to have seen the meteor

(b) the standard deviation of this time, in minutes

(c) the probability that he saw it between 11.12 and 11.15 pm.

Solution Let X denote the number of minutes after 11.00 pm at which the meteor was seen. Then X is uniformly distributed on the interval [0, 15].

(a) $E(X) = \frac{1}{2}(a + b)$
$= \frac{1}{2}(0 + 15)$
$= 7.5$

Thus the mean time is $11.07\frac{1}{2}$ pm.

(b) $\text{Var}(X) = \frac{1}{12}(b - a)^2$
$= \frac{1}{12}(15 - 0)^2$
$= 18.75$

Standard deviation $= \sqrt{\text{Var}(X)} = \sqrt{18.75} = 4.33$ minutes (2 d.p.).

(c) $P(12 \leqslant X \leqslant 15) = \int_{12}^{15} \frac{1}{15}\,dx$
$= \left[\frac{1}{15}x\right]_{12}^{15}$
$= \left[\frac{15}{15} - \frac{12}{15}\right]$
$= 0.2$

EXERCISE 4A

1 The random variable X is uniformly distributed over the interval [0, 20].
(a) Write down $\mathrm{E}(X)$ and $\mathrm{Var}(X)$.
(b) Write down a full description of the p.d.f., $\mathrm{f}(x)$.
(c) Find the probability that X lies between 5 and 12.

2 A guitar string of length 60 cm breaks into two pieces at a random point along its length. The length of the shorter piece is modelled by a uniform continuous variable X with p.d.f. $\mathrm{f}(x)$.
(a) Write down $\mathrm{E}(X)$ and $\mathrm{Var}(X)$.
(b) Write out a full description of the p.d.f., $\mathrm{f}(x)$.
(c) Find the probability that X lies between 0 and 10.

3 The random variable X is uniformly distributed over the interval [0, 1.2].
(a) Write down $\mathrm{E}(X)$.
(b) Use integration to prove that $\mathrm{Var}(X) = 0.12$.
(c) Write down a full description of the p.d.f., $\mathrm{f}(x)$.
(d) Find the probability that X lies between 0.5 and 0.7.
(e) Find the probability that two successive observations of X are both in excess of 0.9.

4 Rosemarie is twirling a conker on a string of length 50 cm, when the string snaps at a random point along its length.
(a) Explain briefly why the length of string remaining attached to the conker may be modelled by a continuous uniform variable, X.
(b) Write down $\mathrm{E}(X)$ and $\mathrm{Var}(X)$.
(c) Find the p.d.f., $\mathrm{f}(x)$.
(d) Find the c.d.f., $\mathrm{F}(x)$.
(e) Find the probability that the length of string which Rosemarie is left holding is less than 10 cm.

5 A robot is placed at the origin of a set of coordinate axes, facing in the direction of the positive x axis. It is programmed to walk a random distance X cm, where X is uniformly distributed between 0 and 50. It then turns left through an angle of 90° and walks forwards a random distance Y cm, where Y is uniformly distributed between 0 and 40. It then stops. The robot will be destroyed if it falls into the rectangular fire pit.

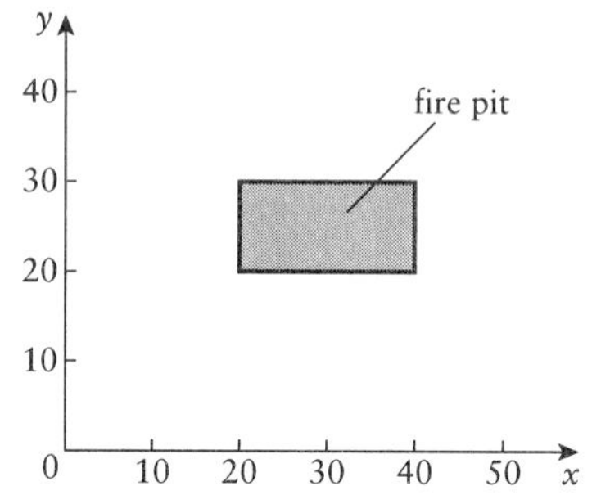

(a) Find the probability that the random variable X lies between 20 and 40.
(b) Find the probability that the robot is destroyed.

MODELLING DISCRETE SITUATIONS

You will recall meeting the normal distribution in *Statistics 1*. Although the normal distribution applies strictly to a continuous variable, it is also common to use it in situations where the variable is discrete providing that:

- the distribution is approximately normal; this requires that the steps in its possible values are small compared with its standard deviation;
- *continuity corrections* are applied where appropriate.

The meaning of the term continuity correction is explained in the following example.

EXAMPLE 4.2

The result of an Intelligence Quotient (IQ) test is an integer score, X. Tests are designed so that X has a mean value of 100 with standard deviation 15. A large number of people have their IQs tested. What proportion of them would you expect to have IQs measuring between 106 and 110 (inclusive)?

Solution

Although the random variable X is an integer and hence discrete, the steps of 1 in its possible values are small compared with the standard deviation of 15. So it is reasonable to treat it as if it is continuous.

If you assume that an IQ test is measuring innate, natural intelligence (rather than the results of learning), then it is reasonable to assume a normal distribution.

If you draw the probability distribution function for the discrete variable X it looks like figure 4.3. The area you require is the total of the five bars representing 106, 107, 108, 109 and 110.

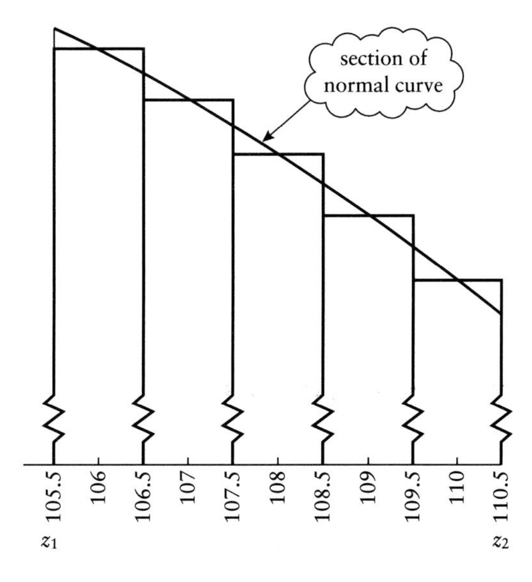

FIGURE 4.3

The equivalent section of the normal curve would run not from 106 to 110 but from 105.5 to 110.5, as you can see in the diagram. When you change from the discrete scale to the continuous scale, the numbers 106, 107 etc. no longer represent the whole intervals, just their centre points.

So the area you require under the normal curve is given by $\Phi(z_2) - \Phi(z_1)$

where $z_1 = \dfrac{105.5 - 100}{15}$ and $z_2 = \dfrac{110.5 - 100}{15}$.

This is $\Phi(0.70) - \Phi(0.37)$
$= 0.7580 - 0.6443 = 0.1137$

Answer: The proportion of IQs between 106 and 110 (inclusive) should be approximately 11%.

In this calculation, both end values needed to be adjusted to allow for the fact that a continuous distribution was being used to approximate a discrete one. These adjustments, $106 \to 105.5$ and $110 \to 110.5$, are called continuity corrections. Whenever a discrete distribution is approximated by a continuous one a continuity correction may need to be used.

You must always think carefully when applying a continuity correction. Should the corrections be added or subtracted? In this case 106 and 110 are inside the required area and so any value (like 105.7 or 110.4) which would round to them must be included. It is often helpful to draw a sketch to illustrate the region you want, like the one in figure 4.3.

If the region of interest is given in terms of inequalities, you should look carefully to see whether they are inclusive ($\leqslant$ or $\geqslant$) or exclusive ($<$ or $>$). For example $20 \leqslant X \leqslant 30$ becomes $19.5 \leqslant X \leqslant 30.5$ whereas $20 < X < 30$ becomes $20.5 \leqslant X \leqslant 29.5$.

Two particularly common situations are when the normal distribution is used to approximate the binomial and the Poisson distributions.

APPROXIMATING THE BINOMIAL DISTRIBUTION

You may use the normal distribution as an approximation for the binomial, $B(n, p)$ (where n is the number of trials each having probability p of success) when

1 n is large
2 p is not too close to 0 or 1.

These conditions ensure that the distribution is reasonably symmetrical and not skewed away from either end, see figure 4.4.

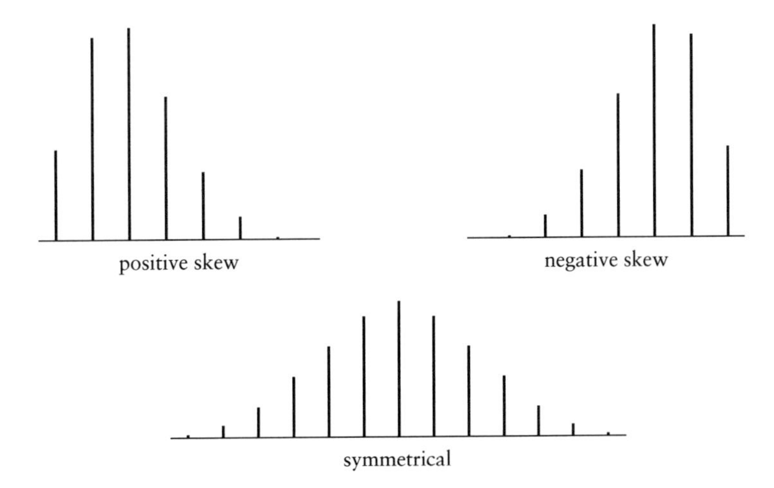

FIGURE 4.4

The parameters for the normal distribution are then

Mean: $\mu = np$

Variance: $\sigma^2 = npq$

so that it can be denoted by $N(np, npq)$.

EXAMPLE 4.3

This is a true story. During voting at a by-election, an exit poll of 1700 voters indicated that 50% of people had voted for the Labour party candidate. When the votes were counted it was found that he had in fact received 57% support.

850 of the 1700 people interviewed said they had voted Labour but 57% of 1700 is 969, a much higher number. What went wrong? Is it possible to be so far out just by being unlucky and asking the wrong people?

Solution The situation of selecting a sample of 1700 people and asking them if they voted for one party or not is one that is modelled by the binomial distribution, in this case $B(1700, 0.57)$.

In theory you could multiply out $(0.43 + 0.57t)^{1700}$ and use that to find the probabilities of getting $0, 1, 2, \ldots, 850$ Labour voters in your sample of 1700. In practice such a method would be impractical because of the work involved.

What you can do is to use a normal approximation. The required conditions are fulfilled: at 1700, n is certainly not small; $p = 0.57$ is near neither 0 nor 1.

The parameters for the normal approximation are given by

Mean, $\mu = np = 1700 \times 0.57 = 969$

S.D., $\sigma = \sqrt{npq} = \sqrt{1700 \times 0.57 \times 0.43} = 20.4$

You will see that the standard deviation, 20.4, is large compared with the steps of 1 in the possible values of Labour voters.

The probability of getting no more than 850 Labour voters, $P(X \leqslant 850)$, is given by $\Phi(z)$, where

$$z = \frac{850.5 - 969}{20.4} = -5.8$$

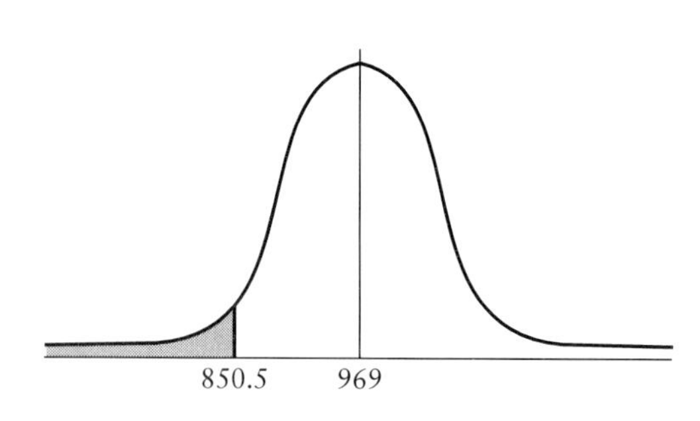

FIGURE 4.5

(Notice the continuity correction making 850 into 850.5.)

This is beyond the range of most tables and corresponds to a probability of about 0.000 01. The probability of a result as extreme as this is thus 0.000 02 (allowing for an equivalent result in the tail above the mean). It is clearly so unlikely that this was a result of random sampling that another explanation must be found.

EXAMPLE 4.4

The random variable X is binomially distributed with $n = 50$ and $p = 0.3$.

(a) Explain briefly why a normal approximation is valid in this particular case.
(b) Write down the parameters of the approximating normal distribution.
(c) Use your approximating distribution to find the probability that X lies between 9 and 20 inclusive. Give your answer correct to 3 decimal places.

Solution

(a) Let $X \sim B(50, 0.3)$.
Since $n\,(50)$ is large and $p\,(0.3)$ is not too close to 0 or 1, the normal approximation is appropriate.

(b) $\mu = np = 50 \times 0.3 = 15$
$\sigma^2 = npq = 50 \times 0.3 \times 0.7 = 10.5$

(c) Using N(15, 10.5),

$$P(9 \leqslant X \leqslant 20) = \Phi\left(\frac{20.5 - 15}{\sqrt{10.5}}\right) - \Phi\left(\frac{8.5 - 15}{\sqrt{10.5}}\right)$$
$$= \Phi(1.70) - \Phi(-2.006)$$
$$= 0.9554 - (1 - 0.9772)$$
$$= 0.9554 - 0.0228$$
$$= 0.9326$$
$$= 0.933 \text{ (3 d.p.)}$$

Since only 2.00 and 2.02 are tabulated, the figure for 2.00 is used.

EXERCISE 4B

1 The intelligence of an individual is frequently described by a positive integer known as an IQ (intelligence quotient). The distribution of IQs amongst children of a certain age-group can be approximated by a normal probability model with mean 100 and standard deviation 15. Write a sentence stating what you understand about the age-group from the fact that $\Phi(2.5) = 0.994$.

A class of 30 children is selected at random from the age-group. Calculate (to 3 significant figures) the probability that at least one member of the class has an IQ of 138 or more.

[SMP]

2 A certain examination has a mean mark of 100 and a standard deviation of 15. The marks can be assumed to be normally distributed.

(a) What is the least mark needed to be in the top 35% of pupils taking this examination?

(b) Between which two marks will the middle 90% of the pupils lie?

(c) 150 pupils take this examination. Calculate the number of pupils likely to score 110 or over.

[MEI]

3 25% of Flapper Fish have red spots, the rest blue spots. A fisherman nets 10 Flapper Fish. What are the probabilities that

(a) exactly 8 have blue spots

(b) at least 8 have blue spots?

A large number of samples, each of 100 Flapper Fish, are taken.

(c) What is

(i) the mean

(ii) the standard deviation of the number of red-spotted fish per sample?

(d) What is the probability of a sample of 100 Flapper Fish containing over 30 with red spots?

4 A fair coin is tossed 10 times. Evaluate the probability that exactly half of the tosses result in heads.

The same coin is tossed 100 times. Use the normal approximation to the binomial to estimate the probability that exactly half of the tosses result in heads. Also estimate the probability that more than 60 of the tosses result in heads.

Explain why a continuity correction is made when using the normal approximation to the binomial and the reason for the adoption of this correction.

[MEI]

5 State conditions under which a binomial probability model can be well approximated by a normal model.

X is a random variable with the distribution B(12, 0.42).

(a) Anne uses the binomial distribution to calculate the probability that $X < 4$ and gives 4 significant figures in her answer. What answer should she get?

(b) Ben uses a normal distribution to calculate an approximation for the probability that $X < 4$ and gives 4 significant figures in his answer. What answer should he get?

(c) Given that Ben's working is correct, calculate the percentage error in his answer.

[Cambridge]

6 During an advertising campaign, the manufacturers of Wolfitt (a dog food) claimed that 60% of dog owners preferred to buy Wolfitt.

(a) Assuming that the manufacturer's claim is correct for the population of dog owners, calculate

(i) using the binomial distribution

(ii) using a normal approximation to the binomial

the probability that at least 6 of a random sample of 8 dog owners prefer to buy Wolfitt. Comment on the agreement, or disagreement, between your two values. Would the agreement be better or worse if the proportion had been 80% instead of 60%?

(b) Continuing to assume that the manufacturer's figure of 60% is correct, use the normal approximation to the binomial to estimate the probability that, of a random sample of 100 dog owners, the number preferring Wolfitt is between 60 and 70 inclusive.

[MEI]

7 A multiple-choice examination consists of 20 questions, for each of which the candidate is required to tick as correct one of three possible answers. Exactly one answer to each question is correct. A correct answer gets 1 mark and a wrong answer gets 0 marks. Consider a candidate who has complete ignorance about every question and therefore ticks at random. What is the probability that he gets a particular answer correct? Calculate the mean and variance of the number of questions he answers correctly.

The examiners wish to ensure that no more than 1% of completely ignorant candidates pass the examination. Use the normal approximation to the binomial, working throughout to 3 decimal places, to establish the pass mark that meets this requirement.

[MEI]

8 A large box contains many plastic syringes, but previous experience indicates that 10% of the syringes in the box are defective. Five syringes are taken at random from the box. Use a binomial model to calculate, giving your answers correct to 3 decimal places, the probability that

(a) none of the five syringes is defective

(b) at least two syringes out of the five are defective.

Discuss the validity of the binomial model in this context.

Instead of removing five syringes, 100 syringes are picked at random and removed. A normal distribution may be used to estimate the probability that at least 15 out of the 100 syringes are defective. Give a reason why it may be convenient to use a normal distribution to do this, and calculate the required estimate.

[Cambridge]

APPROXIMATING THE POISSON DISTRIBUTION

You may use the normal distribution as an approximation for the Poisson distribution, provided that its parameter (mean) λ is sufficiently large for the distribution to be reasonably symmetrical and not positively skewed.

As a working rule λ should be at least 10.

If $\lambda = 10$, mean $= 10$

and standard deviation $= \sqrt{10} = 3.16$.

A normal distribution is almost entirely contained within 3 standard deviations of its mean and in this case the value 0 is slightly more than 3 standard deviations away from the mean value of 10.

The parameters for the normal distribution are then

Mean: $\mu = \lambda$

Variance: $\sigma^2 = \lambda$

so that it can be denoted by $\mathrm{N}(\lambda, \lambda)$.

(Remember that, for a Poisson distribution, mean = variance.)

For values of λ larger than 10 the Poisson probability graph becomes less positively skewed and more bell-shaped in appearance thus making the normal approximation appropriate. Figure 4.6 shows the Poisson probability graph for the two cases $\lambda = 3$ and $\lambda = 25$. You will see that for $\lambda = 3$ the graph is positively skewed but for $\lambda = 25$ it is approximately bell-shaped.

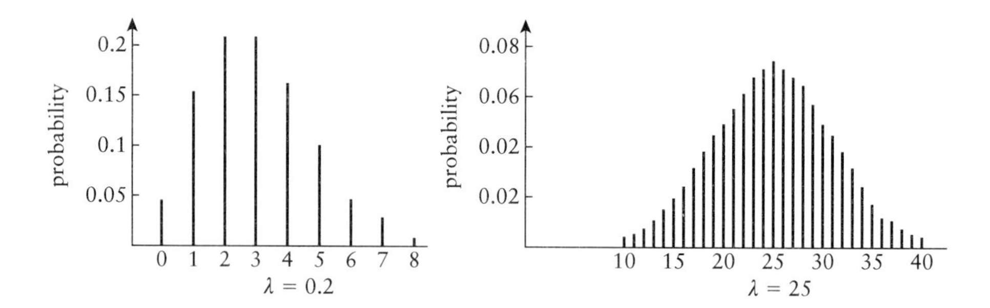

FIGURE 4.6

EXAMPLE 4.5

The annual number of deaths nationally from a rare disease, X, may be modelled by the Poisson distribution with mean 25. One year there are 31 deaths and it is suggested that the disease is on the increase.

What is the probability of 31 or more deaths in a year, assuming the mean has remained at 25?

Solution The Poisson distribution with mean 25 may be approximated by the normal distribution with parameters

Mean: 25

S.D.: $\sqrt{25} = 5.$

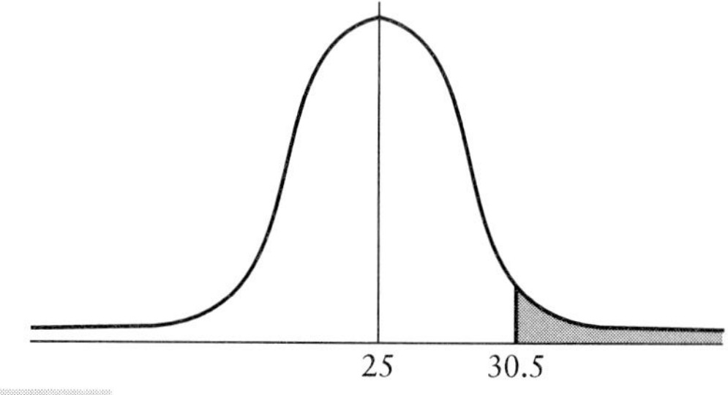

FIGURE 4.7

The probability of there being 31 or more deaths in a year, $P(X \geqslant 31)$, is given by $1 - \Phi(z)$, where

$$z = \frac{30.5 - 25}{5} = 1.1$$

(Note the continuity correction, replacing 31 by 30.5.)

The required area is $1 - \Phi(1.1)$

$$= 1 - 0.8643$$
$$= 0.1357$$

This is not a particularly low probability; it is quite likely that there would be that many deaths in any one year.

EXAMPLE 4.6

The number of first aid kits demanded by customers of a small chemist shop each month is Poisson distributed with parameter 12.5. The chemist restocks his shop at the beginning of each month.

Find the smallest number, k, of first aid kits that he should stock at the beginning of each month so that he has at least a 95% chance of meeting that month's demand.

Solution Let X be the number of kits demanded in a month.
Then $X \sim \text{Po}(12.5)$ and, since this parameter is large, you may use the normal approximation $X \sim \text{N}(12.5, 12.5)$.

Let $$\text{P}(X \geqslant k) = 0.05$$

$$\therefore \quad \Phi\left(\frac{k + 0.5 - 12.5}{\sqrt{12.5}}\right) = 0.05$$

Note the continuity correction

$$\frac{k - 12}{\sqrt{12.5}} = 1.6449$$

$$\therefore \quad k - 12 = 5.816$$

$$k = 17.816$$

$$\therefore \quad k = 18$$

The chemist should stock at least 18 first aid kits.

EXERCISE 4C

1 A telephone exchange serves 2000 subscribers, and at any moment during the busiest period there is a probability of $\frac{1}{30}$ for each subscriber that he will require a line. Assuming that the needs of subscribers are independent, write down an expression for the probability that exactly N lines will be occupied at any moment during the busiest period.

Use the normal distribution to estimate the minimum number of lines that would ensure that the probability that a call cannot be made because all the lines are occupied is less than 0.01.

Investigate whether the total number of lines needed would be reduced if the subscribers were split into two groups of 1000, each with its own set of lines.

[MEI]

2 The number of cars per minute entering a multi-storey car park can be modelled by a Poisson distribution with mean 2. What is the probability that three cars enter during a period of one minute?

What are the mean and the standard deviation of the number of cars entering the car park during a period of 30 minutes? Use the normal approximation to the Poisson distribution to estimate the probability that at least 50 cars enter in any one 30-minute period.

[MEI]

3 State the mean and variance of the Poisson distribution. State under what circumstances the normal distribution can be used as an approximation to the Poisson distribution.

Readings, on a counter, of the number of particles emitted from a radioactive source in a time T seconds have a Poisson distribution with mean $250T$. A ten-second count is made. Find the probabilities of readings of
(a) more than 2600 **(b)** 2400 or more.

[JMB]

4 A drug manufacturer claims that a certain drug cures a blood disease on average 80% of the time. To check the claim, an independent tester uses the drug on a random sample of n patients. He decides to accept the claim if k or more patients are cured.

Assume that the manufacturer's claim is true.

(a) State the distribution of X, the number of patients cured. Find the probability that the claim will be accepted when 15 individuals are tested and k is set at 10.

A more extensive trial is now undertaken on a random sample of 100 patients.

(b) State a suitable approximating distribution for X, and so estimate the probability that the claim will be rejected if k is set at 75.

(c) Find the largest value of k such that the probability of the claim being rejected is no more than 1%.

[MEI]

5 A large computer system which is in constant operation requires an average of 30 service calls per year.

(a) State the average number of service calls per month, taking a month to be $\frac{1}{12}$ of a year. What assumptions need to be made for the Poisson distribution to be used to model the number of calls in a given month?

(b) Use the Poisson distribution to find the probability that at least one service call is required in January. Obtain the probability that there is at least one service call in each month of the year.

(c) The service contract offers a discount if the number of service calls in the year is 24 or fewer. Use a suitable approximating distribution to find the probability of obtaining the discount in any particular year.

[MEI]

6 A national car-hire company has records to show that its vehicles are involved in road traffic accidents at random and at an average rate of 1 per 15 000 miles. The company keeps each car until it has covered 10 000 miles.

(a) Calculate the probability that a car, chosen at random, is involved in
(i) zero **(ii)** more than one
road traffic accidents during the time the company owns it.

The company has a total of 400 cars covering an average of 2000 miles per month each.

(b) Write down the distribution of the number of accidents per month involving the company's cars. Give also a suitable approximating distribution.

(c) Find values a and b such that the number of accidents involving the company's cars in a month is 95% certain to lie between a and b.

[MEI]

EXERCISE 4D

Examination-style questions

1 The random variable X is binomially distributed with $n = 60$ and $p = 0.4$.

(a) Use the binomial distribution to calculate the probability that $X = 24$.

(b) Write down the parameters of a suitable approximating normal approximation.

(c) Use your approximating distribution to find the probability that $X = 24$.

(d) Comment on the agreement between your answers to parts **(a)** and **(c)**.

2 The number of cars passing a certain point per minute, X, may be modelled by a Poisson distribution with parameter 2.5.

(a) Find the probability that less than three cars pass the survey point in a given minute.

(b) Write down the distribution of the number of cars, Y, passing the survey point per hour. Give also a suitable approximating distribution.

(c) Use your approximating distribution to find the probability that the number of cars passing the survey point in a given hour is less than 140.

3 A piece of string AB has length 12 cm. A child cuts the string at a randomly chosen point, P, into two pieces. The random variable X represents the length, in cm, of the piece AP.

(a) Suggest a suitable model for the distribution of X and specify it fully.

(b) Find the cumulative distribution function of X.

(c) Write down $P(X < 4)$.

[Edexcel]

4 Records are kept of the number of road accidents per day on an urban motorway. The data for a random sample of 80 weekdays are summarised by

$$\sum x = 268, \qquad \sum x^2 = 1160.$$

(a) Find the mean and variance of the data.

(b) Give *two* reasons why the Poisson distribution might be thought to be a suitable model for the number of accidents per weekday.

(c) Use the Poisson distribution, with mean as found in part **(a)**, to calculate the expected number of days in a period of 80 weekdays on which there will be exactly one accident.

(d) Calculate the probability that in any period of two particular weekdays there will be a total of more than five accidents.

(e) Using a suitable approximating distribution, calculate the probability of at least 60 accidents in a period of 15 weekdays.

(f) During a particular period of 15 weekdays, roadworks were in progress on the motorway. In this period 60 accidents were reported. Discuss whether this suggests that the number of accidents is abnormally high when roadworks are in progress.

[MEI]

5 A survey conducted by a local education authority requires schools to complete a questionnaire about a sample of their pupils. The sample is defined as all pupils born on the fifth of any month, so that the probability of a randomly chosen pupil being in the sample is $\frac{12}{365}$, or about 0.0329.

The A level year in a secondary school consists of 100 pupils.

(a) Find the probability that none of these pupils appears in the sample.

The school has 1500 pupils in total.

(b) Write down the expectation and the standard deviation of the total number of pupils sampled in a school of this size.

(c) Use a suitable approximating distribution to find the probability that the number of pupils sampled is more than 60.

(d) Find the greatest value, k, such that it is 95% certain that k or more pupils will be sampled.

(e) What should the local authority's reaction be if the school returns just 32 questionnaires?

[MEI]

6 A test consists of 100 multiple choice questions, each having four possible answers labelled A, B, C, D.

Anna does not know the answers to any of the questions so she guesses at random. Find the probability that

(a) she gets none of the first 10 answers right

(b) she gets 4 or more of the first 10 answers right.

(c) Using a suitable approximating distribution, find the probability that Anna gets exactly 25 answers correct on the whole paper.

(d) The pass mark on the paper is 40. Show that Anna is extremely unlikely to pass.

(e) Bella knows the correct answers to 25 questions but guesses at the rest. Obtain Bella's expected score and explain carefully whether or not she is likely to pass.

[MEI]

7 (a) Give two conditions which must apply when modelling a random variable by a Poisson distribution.

A particular make of kettle is sold by a shop at an average rate of five per week. The random variable X represents the number of kettles sold in any one week and X is modelled by a Poisson distribution.

The shop manager notices that at the beginning of a particular week there are seven kettles in stock.

(b) Find the probability that the shop will not be able to meet all the demands for kettles that week, assuming that it is not possible to restock during the week.

In order to increase sales performance, the manager decides to have in stock at the beginning of each week sufficient kettles to have at least a 99% chance of being able to meet all demands during that week.

(c) Find the smallest number of kettles that should be in stock at the beginning of each week.

(d) Using a suitable approximation find the probability that the shop sells at least 18 kettles in a four-week period, subject to stock always being available to meet demand.

[Edexcel]

8 Pak-a-Bik manufactures biscuits which are packed at random in presentation boxes, each box containing 20 biscuits. The company produces 45% chocolate biscuits, the remainder being plain biscuits. Five per cent of all the biscuits made are wrapped in coloured foil.

A box is selected at random from the production line. The random variable C represents the number of chocolate biscuits contained in this box.

(a) Write down two reasons to support the use of the binomial distribution as a suitable model for the random variable C.

(b) Calculate the probability that this box contains

(i) exactly eight chocolate biscuits

(ii) more chocolate biscuits than plain biscuits.

The company quality assurance manager takes a random sample of ten boxes of biscuits.

(c) Find the probability that exactly four of them contain more chocolate biscuits than plain ones.

For a special order, Pak-a-Bik produces a box containing 120 biscuits.

(d) Using suitable approximations, calculate the probability that this box contains

(i) exactly 12 biscuits wrapped in coloured foil

(ii) at least 50 but not more than 60 chocolate biscuits.

[Edexcel]

9 At a children's party each child was blindfolded and asked to pin a tail on a cardboard donkey. The distance, in cm, of the pin from the correct position was measured and the results are recorded below

$$17, \quad 15, \quad 5, \quad 9, \quad 13, \quad 42, \quad 8, \quad 24, \quad 34, \quad 38, \quad 29, \quad 6.$$

(a) Find the mean and the standard deviation for this set of numbers.

A statistics student, who was helping at this party, attempts to model the distance, in cm, a child places the pin from the correct position using the continuous uniform distribution over the interval [0, 50].

(b) Use the formulae in the formula booklet to find the mean and the standard deviation of this distribution.

(c) Comment on the suitability of this distribution as a model in the present situation.

The student attempts to refine the model and considers two distributions with probability density functions $f(x)$ and $g(x)$ illustrated below.

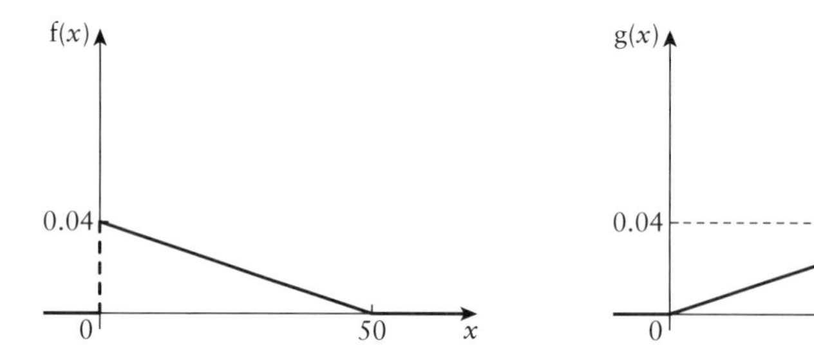

(d) Explain, giving a reason, which of these two probability density functions the student should choose.

[Edexcel]

10 A biologist is studying the behaviour of sheep in a large field. The field is divided into a number of equally sized squares and the average number of sheep per square is 2.5. The sheep are randomly spread throughout the field.

(a) Suggest a suitable model for the number of sheep in a square and give a suitable value for any parameter or parameters required.

Calculate the probability that a randomly selected square contains

(b) no sheep

(c) more than four sheep.

A sheep dog has been sent into the field to round up the sheep.

(d) Explain why the model may no longer be applicable.

In another field the average number of sheep per square is 20 and the sheep are scattered randomly throughout the field.

(e) Using a suitable approximation, find the probability that a randomly selected square contains fewer than 15 sheep.

[Edexcel]

KEY POINTS

1 The continuous uniform distribution over the interval $[a, b]$ has the following properties:

- p.d.f.

$$f(x) = \begin{cases} \dfrac{1}{b-a} & a \leqslant x \leqslant b \\ 0 & \text{otherwise.} \end{cases}$$

- c.d.f.

$$F(x) = \begin{cases} 0 & x < a \\ \dfrac{x-a}{b-a} & a \leqslant x \leqslant b \\ 1 & x > b \end{cases}$$

- mean $E(X) = \dfrac{a+b}{2}$

- variance $\text{Var}(X) = \dfrac{(b-a)^2}{12}$

2 The binomial distribution $B(n, p)$ may be approximated by the normal distribution $N(np, npq)$ provided n is large and p is not too close to 0 or 1. A continuity correction must be used.

3 The Poisson distribution $\text{Po}(\lambda)$ may be approximated by the normal distribution $N(\lambda, \lambda)$ provided λ is reasonably large. A continuity correction must be used.

Chapter five

HYPOTHESIS TESTS

You may prove anything by figures.

An anonymous salesman

SAMPLING

Suppose you want to investigate the number of adult smokers in the UK. There are two quite different approaches to this problem: you can conduct a census, i.e. ask everyone in the UK to take part in the investigation or you can conduct a sample survey, i.e. select a group of people from within the UK and then ask them to take part in the investigation.

Each of these approaches has its own advantages and disadvantages. The census has the advantage that the data set is complete, but it can be very difficult, if not impossible, to obtain all the necessary data. The sample survey makes the data easier to collect and process, but it may distort the picture unless carried out carefully.

Taking samples and interpreting them comprises an essential part of statistics. The populations encountered are often so large that it would be quite impractical to use every item.

A *sample* provides a set of data values of a random variable, drawn from all such possible values, the *parent population*. The parent population can be finite, such as all professional footballers, or infinite, such as the points where a dart can land on a dart board.

The individuals available for sampling are called the *sampling units*. A representation of the items available to be sampled is called the *sampling frame*. This could, for example, be a list of the sheep in a flock, a map marked with a grid or an electoral register. In many situations no sampling frame exists nor is it possible to devise one, for example, for the cod in the North Atlantic. The proportion of the available items that are actually sampled is called the *sampling fraction*.

A parent population, often just called the *population*, is described in terms of its *parameters*, such as its mean, μ, and variance, σ^2. By convention Greek letters are used to denote these parameters.

A value derived from a sample is written in Roman letters: mean, $\overline{x}$, variance, s^2, etc. Such a number is the value of a *sample statistic* (or just *statistic*). When sample statistics are used to estimate the parent population parameters they are called *estimates*.

Thus if you take a random sample in which the mean is $\overline{x}$, you can use $\overline{x}$ to estimate the parent mean, μ. If in a particular sample $\overline{x} = 23.4$, then you can use 23.4 as an estimate of the population mean. The true value of μ will generally be somewhat different from your estimated value.

Upper case letters, X, Y, etc., are used to represent the random variables, and lower case letters, x, y, etc., to denote particular values of them.

There are essentially two reasons why you might wish to take a sample.

- To estimate the values of the parameters of the parent population.
- To conduct a hypothesis test.

There are many ways you can interpret data. First you will consider how sample data are collected and the steps you can take to ensure their quality.

An estimate of a parameter derived from sample data will in general differ from its true value. The difference is called the *sampling error*. To reduce the sampling error, you want your sample to be as representative of the parent population as you can make it. This, however, may be easier said than done.

Here are a number of questions that you should ask yourself when about to take a sample.

1 Are the data relevant?

It is a common mistake to replace what you need to measure by something else for which data are more easily obtained.

You must ensure that your data are relevant, giving values of whatever it is that you really want to measure.

2 Are the data likely to be biased?

Bias is a systematic error. If, for example, you wished to estimate the mean time of young women running 100 metres and did so by timing the members of a hockey team over that distance, your result would be biased. The hockey players would be fitter and more athletic than most young women and so your estimate for the time would be too low.

You must try to avoid bias in the selection of your sample.

3 Does the method of collection distort the data?

The process of collecting data must not interfere with the data. It is, for example, very easy when designing a questionnaire to frame questions in such a way as to lead people into making certain responses. 'Are you a law-abiding citizen?' and 'Do you consider your driving to be above average?' are both questions inviting the answer 'Yes'.

In the case of collecting information on voting intentions another problem arises. Where people put the cross on their ballot papers is secret and so people are being asked to give away private information. There may well be those who find this offensive and react by deliberately giving false answers.

People often give the answer they think the questioner wants to receive.

4 Is the right person collecting the data?

Bias can be introduced by the choice of those taking the sample. For example, a school's authorities want to estimate the proportion of the students who smoke, which is against the school rules. Each form teacher is told to ask five students whether they smoke. Almost certainly some smokers will say 'No' to their teacher for fear of getting into trouble, even though they might say 'Yes' to a different person.

5 Is the sample large enough?

The sample must be sufficiently large for the results to have some meaning. For opinion polls, a sample size of about 1000 is common.

The sample size depends on the precision required in the results. For example, in the opinion polls for elections a much larger sample is required if you want the estimate to be reliable to within 1% than if 5% will do.

6 Is the sampling procedure appropriate in the circumstances?

The method of choosing the sample must be appropriate. Suppose, for example, that you were carrying out a survey of people's voting intentions in a forthcoming by-election. How would you select the sample of people you are going to ask?

If you stood in the town's high street in the middle of one morning and asked passers-by you would probably get an unduly high proportion of those who, for one reason or another, were not employed. It is quite possible that this group has different voting intentions from those in work.

If you selected names from the telephone directory, you would automatically exclude those who do not have telephones: the lower income groups, students and so on.

It is actually very difficult to come up with a plan which will yield a fair sample, one that is not biased in some direction or another. There are, however, a number of established sampling techniques and these are described in the next section of this chapter.

SAMPLING TECHNIQUES

In considering the following techniques it is worth repeating that a key aim when taking a sample is to obtain a sample that is *representative* of the parent population being investigated. It is assumed that the sampling is done without replacement, otherwise, for example, one person could give an opinion twice, or more. The fraction of the population which is selected is called the *sampling fraction*.

$$\text{Sampling fraction} = \frac{\text{sample size}}{\text{population size}}$$

SIMPLE RANDOM SAMPLING

In a *simple random sampling procedure*, every possible sample of a given size is equally likely to be selected. It follows that in such a procedure every member of the parent population is equally likely to be selected. However, the converse is not true. It is possible to devise a sampling procedure in which every member is equally likely to be selected but some samples are not permissible.

Simple random sampling is fine when you can do it, but you must have a sampling frame. The selection of items within the frame is often done using tables of random numbers. Random numbers can be generated using a calculator or computer program.

STRATIFIED SAMPLING

You have already thought about the difficulty of conducting a survey of people's voting intentions in a particular area before an election. In that situation it is possible to identify a number of different sub-groups which you might expect to have different voting patterns: low, medium and high income groups; urban, suburban and rural dwellers; young, middle-aged and elderly voters; men and women; and so on. The sub-groups are called *strata*. In *stratified sampling*, you would ensure that all strata were sampled. You would need to sample from high income, suburban, elderly women; medium income, rural young men; etc. In this example, 54 strata ($3 \times 3 \times 3 \times 2$) have been identified. If the numbers sampled in the various strata are proportional to the size of their populations, the procedure is called *proportional stratified sampling*. If the sampling is not proportional, then appropriate weighting has to be used.

The selection of the items to be sampled within each stratum is usually done by simple random sampling. Stratified sampling will usually lead to more accurate results about the entire population, and will also give useful information about the individual strata.

Cluster sampling

Cluster sampling also starts with sub-groups, or strata, of the population, but in this case the items are chosen from one or several of the sub-groups. The sub-groups are now called clusters. It is important that each cluster should be reasonably representative of the entire population. If, for example, you were asked to investigate the incidence of a particular parasite in the puffin population of Northern Europe, it would be impossible to use simple random sampling. Rather you would select a number of sites and then catch some puffins at each place. This is cluster sampling. Instead of selecting from the whole population you are choosing from a limited number of clusters.

Systematic sampling

Systematic sampling is a method of choosing individuals from a sampling frame. If you were surveying telephone subscribers, you might select a number at random, say 66, and then sample the 66th name on every page of the directory. If the items in the sampling frame are numbered $1, 2, 3, \ldots$, you might choose a random starting point like 38 and then sample numbers 38, 138, 238 and so on.

When using systematic sampling you have to beware of any cyclic patterns within the frame. For example, a school list is made up form by form, each of exactly 25 children, in order of merit, so that numbers $1, 26, 51, 76, 101, \ldots$, in the frame are those at the top of their forms. If you sample every 50th child starting with number 26, you will conclude that the children in the school are very bright.

Quota sampling

Quota sampling is the method often used by companies employing people to carry out opinion surveys. An interviewer's quota is always specified in stratified terms, how many males and how many females, etc. The choice of who is sampled is then left up to the interviewer and so is definitely non-random.

Other sampling techniques

This is by no means a complete list of sampling techniques. *Survey design*, the formulation of the most appropriate sampling procedure in a particular situation, is a major topic within statistics.

EXERCISE 5A

1 (a) An accountant is sampling from a computer file. The first number is selected randomly and is item 47; the rest of the sample is selected automatically and comprises items 97, 147, 197, 247, 297,....

What type of sampling procedure is being used?

(b) Pritam is a pupil at a local school. He has been given a copy of the list of all students in the school. The list numbers the students from 1 to 2500.

Pritam generates a four-digit random number on his calculator, for example 0.4325. He multiplies the random number by 2500 and notes the integer part. For example, 0.4325×2500 results in 1081 so Pritam chooses the pupil listed as 1081. He repeats the process until he has a sample of 100 names.

(i) What type of sampling procedure is Pritam carrying out?
(ii) What is the sampling fraction in this case?

(c) Mr Jones wishes to find out if a mobile grocery service would be popular in Southley. He chooses four streets at random in the town and calls at 15 randomly selected houses in each of the streets to seek the residents' views.

(i) What type of sampling procedure is he using?
(ii) Is the procedure random?

(d) Tracey is trying to encourage people to shop at her boutique. She has produced a short questionnaire and has employed four college students to administer it. The questionnaire asks people about their fashion preferences. Each student is told to question 20 women and 20 men and then to stop.

(i) What type of sampling procedure is Tracey using?
(ii) Is the procedure random?
(iii) Comment on the number of people that are surveyed.

2 (a) There are five year groups in the school Jane attends. She wishes to survey opinion about what to do with an unused section of field next to the playground. Because of a limited budget she has produced only 30 questionnaires.

There are 140 students in each of Years 1 and 2.
There are 100 students in each of Years 3 and 4.
There are 120 students in Year 5.

Jane plans to use a stratified sampling procedure.

(i) How many students from each year should Jane survey?
(ii) What is the sampling fraction?

(b) A factory safety inspector wishes to inspect a sample of vehicles to check for faulty tyres. The factory has 280 light vans, 21 company cars and 5 large-load vehicles.

The chairman has instructed that a sampling fraction of $\frac{1}{10}$ should be used and that each type of vehicle should be represented in the sample.

(i) How many of each vehicle type should be inspected?

(ii) How should the inspector choose his sample? What is the sampling procedure called?

(c) A small village has a population of 640. The population is classified by age as shown in the table below.

Age (years)	0–5	6–12	13–21	22–35	36–50	51+
Number of people	38	82	108	204	180	28

A survey of the inhabitants of the village is intended. A sample of size 80 is proposed.

(i) What is the overall sampling fraction?

(ii) A stratified sample is planned. Calculate the approximate number that should be sampled from each age group.

3 Identify the sampling procedures that would be appropriate in the following situations.

(a) A local education officer wishes to estimate the mean number of children per family on a large housing estate.

(b) A consumer protection body wishes to estimate the proportion of trains that are running late.

(c) A marketing consultant wishes to investigate the proportion of households in a town that have a personal computer.

(d) A local politician wishes to carry out a survey into people's views on capital punishment within your area.

(e) A health inspector wishes to investigate what proportion of people wear spectacles.

(f) Ministry officials wish to estimate the proportion of cars with bald tyres.

(g) A television company wishes to estimate the proportion of householders who have not paid their television licence fee.

(h) The police want to find out how fast cars travel in the outside lane of a motorway.

(i) A sociologist wants to know how many girlfriends the average 18-year-old boy has had.

(j) The headteacher of a large school wishes to estimate the average number of hours homework done per week by the students.

4 You have been given the job of refurnishing the college canteen. You wish to survey student opinion on this. You are considering a number of sampling methods. In each case describe the sampling method and list the advantages and disadvantages.

(a) Select every 25th student from the college's alphabetical listing of students.

(b) Select students as they arrive at college, ensuring proportional numbers of males and females and from classes on different courses.

(c) Select students as they enter the canteen.

(d) Select students at random from first and second year-group listings and in proportion to the number on each list.

5 Sampling is required in the situations below. For each situation devise, name and describe a suitable strategy. (Your answer is expected to take about five to ten lines for each part.)

(a) A company producing strip lighting wishes to find an estimate of the life expectancy of a typical strip light. Suggest how they might obtain a suitable sample.

(b) A tree surgeon wishes to estimate the number of damaged trees in a large forest. He has available a map of the forest. Suggest how he might select a sample.

(c) A factory produces computer chips. It has five production lines. Each production line produces, on average, 100 000 chips per week. One week the quality control manager decides to take a random sample of 500 chips from each production line.

(i) Describe how she might arrange for a sample to be taken from a production line.

(ii) What sampling method is she employing overall?

(d) The local college is anxious to monitor the use of the college car park, which has parking spaces for 100 cars. It is aware that the number of staff employed by the college is greater than this but also that some staff use public transport sometimes. It is considering giving staff a choice of a parking permit (cost as yet undecided) or paying for staff to use public transport.

How would you survey staff views on these proposals?

(e) Some of your fellow pupils have shown concern about the lack of available space to do private study. You have been asked to represent them in approaching the Principal in order to press for some improvement in appropriate study space. Before you do this you want to be sure that you are representing a majority view, not just the feelings of a few 'complaining' individuals.

Describe how you would survey the pupils to gain the required information.

HYPOTHESIS TEST FOR THE BINOMIAL DISTRIBUTION

Suppose a coin is tossed eight times and the results are eight consecutive heads. Does this indicate that the coin is biased?

To answer this question, you need to consider that the result of eight heads, while unlikely, is still a feasible result when a coin is tossed eight times. The binomial distribution will be used to assess just how unlikely such a result really is.

Suppose the coin is fair. Then the probability p of obtaining a head on any one throw is 0.5. This forms the *null hypothesis*, denoted by H_0:

$$H_0: p = 0.5.$$

On the other hand, suppose the coin is biased in such a way that heads occur more than 50% of the time. Then the probability p of obtaining a head on any one throw is greater than 0.5. This forms the *alternative hypothesis*, denoted by H_1:

$$H_1: p > 0.5.$$

So, if H_0 is true then the number of heads, X, obtained when the coin is tossed eight times may be modelled by the binomial distribution $X \sim B(8, 0.5)$.

Under H_0 the probabilities may be calculated using the binomial model B(8, 0.5) as shown in the table:

Number of heads	Probability	Cumulative probability (%)
0	$\frac{1}{256}$	0.4
1	$\frac{8}{256}$	3.5
2	$\frac{28}{256}$	14.4
3	$\frac{56}{256}$	36.3
4	$\frac{70}{256}$	63.7
5	$\frac{56}{256}$	85.5
6	$\frac{28}{256}$	96.5
7	$\frac{8}{256}$	99.6
8	$\frac{1}{256}$	100

To carry out the test at the 5% *significance level* we separate the distribution above into two parts. The lower 95% forms the *non-critical region* while the upper 5% forms the *critical region*. (Since it is not possible to hit 95% exactly, the non-critical region extends until 95% has been exceeded.)

Number of heads	Probability	Cumulative probability (%)
0	$\frac{1}{256}$	0.4
1	$\frac{8}{256}$	3.5
2	$\frac{28}{256}$	14.4
3	$\frac{56}{256}$	36.3
4	$\frac{70}{256}$	63.7
5	$\frac{56}{256}$	85.5
6	$\frac{28}{256}$	96.5
7	$\frac{8}{256}$	99.6
8	$\frac{1}{256}$	100

0 to 6 comprise the non-critical region. This sort of behaviour should occur over 95% of the time

7 and 8 comprise the critical region

The two regions are used to judge the outcome of the hypothesis test. If the observed number of heads had been 0 to 6 inclusive then this would be judged to be consistent with $p = 0.5$, i.e. you would accept H_0.

If, on the other hand, the observed number of heads was 7 to 8, then you are observing behaviour that should only occur 5% (or less) of the time. You would conclude that this is not the usual behaviour when $p = 0.5$, so you would reject H_0 and accept H_1 instead.

The observed number of heads, in this case eight, is referred to as a *test statistic*, and may be indicated on a diagram of the distribution as shown in figure 5.1.

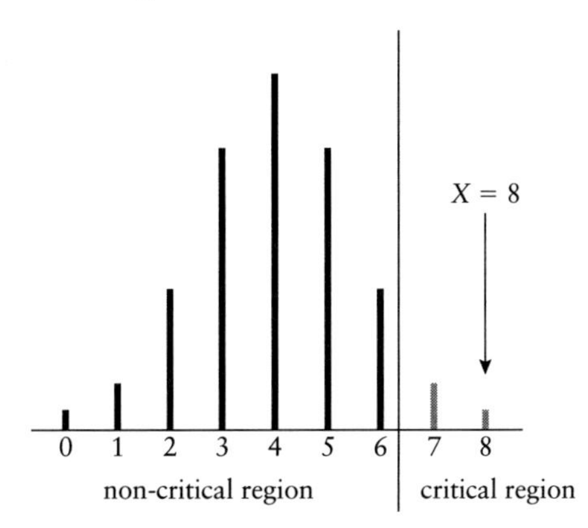

FIGURE 5.1

In this case $X_{\text{test}} = 8$ lies in the critical region.

Therefore you reject H_0 and accept H_1. There does seem to be evidence, at the 5% significance level, that the coin is biased in favour of heads.

ONE-TAIL TESTS

The question about the coins was an example of a *one-tail test*, in which the critical region occupies a tail at one end of the distribution. The tail may occur either at the upper or lower end of the distribution, and the wording of the question must be checked carefully to see which it is.

It is often convenient to use cumulative tables to identify the critical region. Particular care must be taken when working with an upper-tail test. The next two examples emphasise the subtle differences between lower- and upper-tail tests when using these tables.

EXAMPLE 5.1

David has written a computer program which is supposed to generate random whole numbers between 1 and 10 inclusive. He is suspicious that the program does not produce the number 1 as often as it should, so he takes a random sample of 50 numbers generated by the program. The number 1 occurs only twice in the sample.

(a) Write down null and alternative hypotheses which David could test. Explain why the alternative hypothesis takes this particular form.
(b) Find the critical region for a 5% significance test.
(c) Conduct the test, stating your conclusion clearly.

Solution

(a) Let p be the probability of obtaining a number 1.
Then the hypotheses to be tested are:

$$H_0: p = 0.1$$
$$H_1: p < 0.1$$

H_0 takes this form because David is suspicious that the number 1 does not occur often enough, hence a lower-tail test is required.

(b) From tables of B(50, 0.1),

	$p =$	0.05	0.10	0.15	0.20	0.25	0.30	0.35	0.40	0.45	0.50
$n = 50, x =$	0	0.0769	0.0052	0.0003	0.0000	0.0000	0.0000	0.0000	0.0000	0.0000	0.0000
	1	0.2794	0.0338	0.0029	0.0002	0.0000	0.0000	0.0000	0.0000	0.0000	0.0000
	2	0.5405	0.1117	0.0142	0.0013	0.0001	0.0000	0.0000	0.0000	0.0000	0.0000
	3	0.7604	0.2503	0.0460	0.0057	0.0005	0.0000	0.0000	0.0000	0.0000	0.0000

FIGURE 5.2

you can see that the critical region comprises {0, 1} since 3.38% < 5%, whereas 11.17% > 5%.

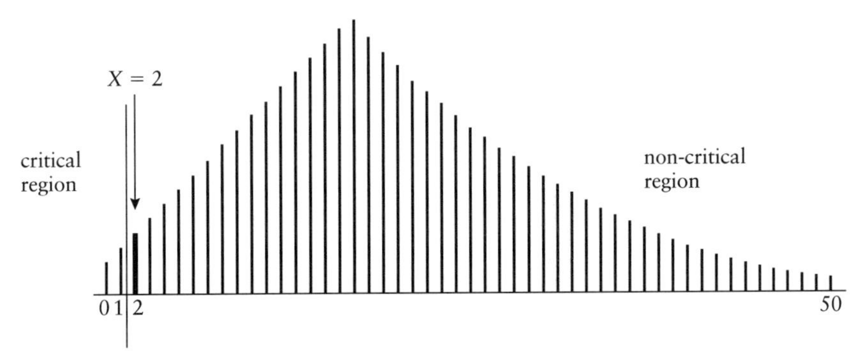

FIGURE 5.3

(c) Since X = 2 is **not** in the critical region then you accept H_0.
There is no evidence that the number of 1s is to low.

Note It is essential to translate the conclusion 'accept H_0' into plain English at the end of the question.

EXAMPLE 5.2

A company is producing silicon chips for use in video cameras. Over a long period of time it has become established that 25% of the chips are of Class 1 quality, i.e. they contain very few defects.

The company has recently improved its production methods, in the hope of increasing the proportion of Class 1 chips. From a random sample of 12 chips, 7 are found to be of Class 1 quality.

(a) Write down null and alternative hypotheses which could be tested. Explain why the alternative hypothesis takes this particular form.
(b) Find the critical region for a 5% significance test.
(c) Conduct the test, stating your conclusion clearly.

Solution **(a)** Let p be the probability of a given chip being of Class 1 quality. Then the hypotheses to be tested are:

$$H_0: p = 0.25$$
$$H_1: p > 0.25$$

H_1 takes this form because the company is hoping to increase the proportion of Class 1 chips, hence an upper-tail test is required.

(b) From tables of B(12, 0.25),

	p = 0.05	0.10	0.15	0.20	0.25	0.30	0.35	0.40	0.45	0.50
$n = 12, x = 0$	0.5404	0.2824	0.1422	0.0687	0.0317	0.0138	0.0057	0.0022	0.0008	0.0002
1	0.8816	0.6590	0.4435	0.2749	0.1584	0.0850	0.0424	0.0196	0.0083	0.0032
2	0.9804	0.8891	0.7358	0.5583	0.3907	0.2528	0.1513	0.0834	0.0421	0.0193
3	0.9978	0.9744	0.9078	0.7946	0.6488	0.4925	0.3467	0.2253	0.1345	0.0730
4	0.9998	0.9957	0.9761	0.9274	0.8424	0.7237	0.5833	0.4382	0.3044	0.1938
5	1.0000	0.9995	0.9954	0.9806	0.9456	0.8822	0.7873	0.6652	0.5269	0.3872
6	1.0000	0.9999	0.9993	0.9961	0.9857	0.9614	0.9154	0.8418	0.7393	0.6128
7	1.0000	1.0000	0.9999	0.9994	0.9972	0.9905	0.9745	0.9427	0.8883	0.8062

Figure 5.4

you can see that the non-critical region comprises{0, 1, ..., 6} since 95% < 98.57%. Thus the critical region is {7, 8, ..., 12}.

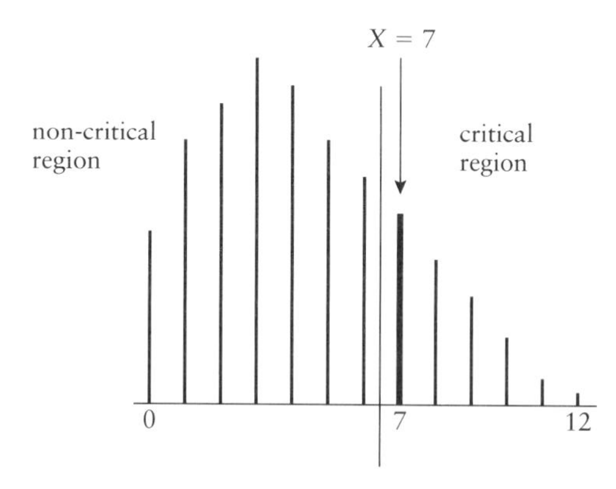

Figure 5.5

(c) Since $X = 7$ is in the critical region you reject H_0 and accept H_1. There is evidence that the proportion of Class 1 chips has increased.

Exercise 5B

1 Mrs da Silva is running for President. She claims to have 60% of the population supporting her.

She is suspected of overestimating her support and a random sample of 20 people are asked whom they support. Only nine say Mrs da Silva.

Test, at the 5% significance level, the hypothesis that she has overestimated her support.

2 A driving instructor claims that 60% of his pupils pass their driving test at the first attempt. Supposing this claim is true, find, using tables or otherwise, the probability that, of 20 pupils taking their first test

(a) exactly 12 pass

(b) more than 12 pass.

A local newspaper reporter suspects that the driving instructor is exaggerating his success rate and so she decides to carry out a statistical investigation. State the null and alternative hypotheses which she is testing. She contacts 20 of his pupils who have recently taken their first test and discovers that N passed. Given that she performs a 5% significance test and that she concludes that the driving instructor's claim was exaggerated, what are the possible values of N?

[MEI]

3 A company developed synthetic coffee and claim that coffee drinkers could not distinguish it from the real product. A number of coffee drinkers challenged the company's claim, saying that the synthetic coffee tasted synthetic. In a test, carried out by an independent consumer protection body, 20 people were given a mug of coffee. Ten had the synthetic brand and ten the natural, but they were not told which they had been given.

Out of the ten given the synthetic brand, eight said it was synthetic and two said it was natural. Use this information to test the coffee drinkers' claim (as against the null hypothesis of the company's claim), at the 5% significance level.

4 A group of 15 students decides to investigate the truth of the saying that if you drop a piece of toast it is more likely to land butter-side down. They each take one piece of toast, butter it on one side and throw it in the air. Eleven land butter-side down, the rest butter-side up. Use their results to carry out a hypothesis test at the 10% significance level, stating clearly your null and alternative hypotheses.

5 On average 70% of people pass their driving test first time. There are complaints that Mr McTaggart is too harsh and so, unknown to himself, his work is monitored. It is found that he fails 10 out of 20 candidates. Are the complaints justified at the 5% significance level?

6 A machine makes bottles. In normal running 5% of the bottles are expected to be cracked, but if the machine needs servicing this proportion will increase. As part of a routine check, 50 bottles are inspected and 5 are found to be unsatisfactory. Does this provide evidence, at the 5% significance level, that the machine needs servicing?

7 A firm producing mugs has a quality control scheme in which a random sample of 10 mugs from each batch is inspected. For 50 such samples, the numbers of defective mugs are as follows:

Number of defective mugs	0	1	2	3	4	5	6+
Number of samples	5	13	15	12	4	1	0

(a) Find the mean and standard deviation of the number of defective mugs per sample.

(b) Show that a reasonable estimate for p, the probability that a mug is defective, is 0.2. Use this figure to calculate the probability that a randomly chosen sample will contain exactly two defective mugs. Comment on the agreement between this value and the observed data.

The management is not satisfied with 20% of mugs being defective and introduces a new process to reduce the proportion of defective mugs.

(c) A random sample of 20 mugs, produced by the new process, contains just one which is defective. Test, at the 5% level, whether it is reasonable to suppose that the proportion of defective mugs has been reduced, stating your null and alternative hypotheses clearly.

(d) What would the conclusion have been if the management had chosen to conduct the test at the 10% level?

[MEI]

8 In a certain country, 90% of letters are delivered the day after posting. A resident posts eight letters on a certain day.

Find the probability that

(a) all eight letters are delivered the next day

(b) at least six letters are delivered the next day

(c) exactly half the letters are delivered the next day.

It is later suspected that the service has deteriorated as a result of mechanisation. To test this, 15 letters are posted and it is found that only 11 of them arrive the next day. Let p denote the probability, after mechanisation, that a letter is delivered the next day.

(d) Write down suitable null and alternative hypotheses for the value of p.

(e) Carry out the hypothesis test, at the 5% level of significance, stating your results clearly.

(f) Write down the critical region for the test, giving a reason for your choice.

[MEI, adapted]

9 For most small birds, the ratio of males to females may be expected to be about 1:1. In one ornithological study birds are trapped by setting fine-mesh nets. The trapped birds are counted and then released. The catch may be regarded as a random sample of the birds in the area.

The ornithologists want to test whether there are more male blackbirds than females.

(a) Assuming that the sex ratio of blackbirds is 1:1, find the probability that a random sample of 16 blackbirds contains

(i) 12 males **(ii)** at least 12 males.

(b) State the null and alternative hypotheses the ornithologists should use.

In one sample of 16 blackbirds there are 12 males and 4 females.

(c) Carry out a suitable test using these data at the 5% significance level, stating your conclusion clearly. Find the critical region for the test.

(d) Another ornithologist points out that, because female birds spend much time sitting on the nest, females are less likely to be caught than males. Explain how this would affect your conclusions.

[MEI]

10 A seed supplier advertises that, on average, 80% of a certain type of seed will germinate. Suppose that 18 of these seeds, chosen at random, are planted.

(a) Find the probability that 17 or more seeds will germinate if

(i) the supplier's claim is correct

(ii) the supplier is incorrect and 82% of the seeds, on average, germinate.

Mr Brewer is the advertising manager for the seed supplier. He thinks that the germination rate may be higher than 80% and he decides to carry out a hypothesis test at the 10% level of significance. He plants 18 seeds.

(b) Write down the null and alternative hypotheses for Mr Brewer's test, explaining why the alternative hypothesis takes the form it does.

(c) Find the critical region for Mr Brewer's test. Explain your reasoning.

(d) Determine the probability that Mr Brewer will reach the *wrong* conclusion if

(i) the true germination rate is 80%

(ii) the true germination rate is 82%.

[MEI]

TWO-TAIL TESTS

In the problems encountered so far you have either tested for a decrease in p (lower-tail test) or for an increase in p (upper-tail test). Both these situations give rise to one-tail tests.

Sometimes, however, you may just want to look for evidence that p has changed, it might have increased or decreased. In such circumstances there will be two critical regions, one at each end of the distribution. This is known as a *two-tail test*. To carry out a two-tail test at the x% significance level you effectively conduct two separate one-tail tests, each at the $\frac{1}{2}x$% significance level.

EXAMPLE 5.3

Pepper moths occur in two varieties, light and dark. The proportion of dark moths increases with certain types of atmospheric pollution.

In a particular village, 25% of the moths are dark, the rest light. A biologist wants to use them as a pollution indicator. She traps samples of 15 moths and counts how many of them are dark.

For what numbers of dark moths among the 15 can she say, at the 10% significance level, that the pollution level is changing?

Solution In this question you are asked to find the critical region for the test:

H_0: $p = 0.25$ (the proportion of dark moths is 25%)
H_1: $p \neq 0.25$ (the proportion is no longer 25%)
Significance level 10%
2-tail test

where p is the probability that a moth selected at random is dark.

You want to find each tail to be as nearly as possible 5% but both must be less than 5%, something that is easiest done using cumulative binomial distribution tables.

Look under $n = 15$, for $p = 0.25$.

p =	0.05	0.10	0.15	0.20	0.25	0.30	0.35	0.40
$n = 15, x = 0$	0.4633	0.2059	0.0874	0.0352	0.0134	0.0047	0.0016	0.0005
1	0.8290	0.5490	0.3186	0.1671	0.0802	0.0353	0.0142	0.0052
2	0.9638	0.8159	0.6042	0.3980	0.2361	0.1268	0.0617	0.0271
3	0.9945	0.9444	0.8227	0.6482	0.4613	0.2969	0.1727	0.0905
4	0.9994	0.9873	0.9383	0.8358	0.6865	0.5155	0.3519	0.2173
5	0.9999	0.9978	0.9832	0.9389	0.8516	0.7216	0.5643	0.4032
6	1.0000	0.9997	0.9964	0.9819	0.9434	0.8689	0.7548	0.6098
7	1.0000	1.0000	0.9994	0.9958	0.9827	0.9500	0.8868	0.7869
8	1.0000	1.0000	0.9999	0.9992	0.9958	0.9848	0.9578	0.9050
9	1.0000	1.0000	1.0000	0.9999	0.9992	0.9963	0.9876	0.9662
10	1.0000	1.0000	1.0000	1.0000	0.9999	0.9993	0.9972	0.9907
11	1.0000	1.0000	1.0000	1.0000	1.0000	0.9999	0.9995	0.9981
12	1.0000	1.0000	1.0000	1.0000	1.0000	1.0000	0.9999	0.9997
13	1.0000	1.0000	1.0000	1.0000	1.0000	1.0000	1.0000	1.0000
14	1.0000	1.0000	1.0000	1.0000	1.0000	1.0000	1.0000	1.0000

FIGURE 5.6

For a two-tail 10% test you need to find two separate 5% critical regions, one at each end.

At the lower end, 0.0134 shows that the left-hand critical region is {0}.

At the upper end, the value of 0.9827, corresponding to $X = 7$, is the first value over 95%, so this marks the end of the non-critical region. The right-hand critical region is therefore {8, 9, . . ., 15}.

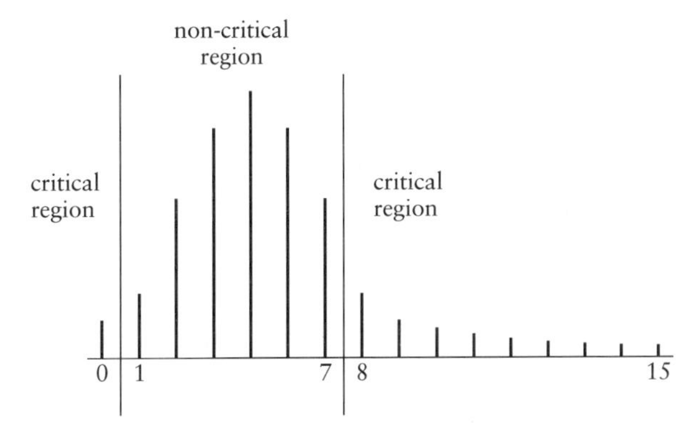

FIGURE 5.7

The overall critical region is thus {0, 8, 9, 10, 11, 12, 13, 14, 15}. For these values she would claim that the pollution level is changing.

EXERCISE 5C

1 To test the claim that a coin is biased, it is tossed 20 times. It comes down heads 12 times. Test at the 10% significance level whether this claim is justified.

2 A biologist discovers a colony of a previously unknown type of bird nesting in a cave. Out of the 15 chicks which hatch during his period of investigation, 13 are female. Test at the 5% significance level whether this supports the view that the sex ratio for the chicks differs from 1:1.

3 People entering an exhibition have to choose whether to turn left or right. Out of the first twelve people, nine turn left and three right. Test at the 5% significance level whether people are more likely to turn one way than another.

4 Weather records for a certain seaside resort show that on average one day in four in April is wet, but local people write to their newspaper complaining that the climate is changing.

A reporter on the paper records the weather for the next 20 days in April and finds that 10 of them are wet.

Do you think the complaint is justified?

5 In a fruit machine there are five drums which rotate independently to show one out of five types of fruit each (lemon, apple, orange, melon and pear). You win a prize if all five stop showing the same fruit. A customer claims that the machine is fixed; the lemon in the first place is not showing the right number of times. The manager runs the machine 20 times and the lemon shows 7 times in the first place. Is the customer's complaint justified at the 10% significance level?

6 A boy is losing in a game of cards and claims that his opponent is cheating.

Out of the last 12 times he shuffled and dealt the cards, the first card to be laid down was a spade on only one occasion. Can he justify his claim at the 2% significance level?

7 A small colony of 20 individuals of a previously unknown animal is discovered. It is believed it might be the same species as one described by early explorers who said that one-quarter of them were male, the rest female.

What numbers of males and females would lead you to believe, at the 5% significance level, that they are not the same species?

8 A multiple choice test has 20 questions, with the answer for each allowing four options, A, B, C and D. All the students in a class tell their teacher that they guessed all 20 answers. The teacher does not believe them. Devise a 2-tail test at the 10% significance level to apply to a student's mark to test the hypothesis that the answers were not selected at random.

9 When a certain language is written down, 15% of the letters are Z. Use this information to devise a test at the 10% significance level which somebody who does not know the language could apply to a short passage, 50 letters long, to determine whether it is written in the same language.

10 A seed firm states on the packets of bean seeds that the germination rate is 80%. Each packet contains 25 seeds.

(a) How many seeds would you expect to germinate out of one packet?
(b) What is the probability of exactly 17 germinating?

A man buys a packet and only 12 germinate.

(c) Is he justified in complaining?

11 Given that X has a binomial distribution in which $n = 15$ and $p = 0.5$, find the probability of each of the following events.

(a) $X = 4$
(b) $X \leqslant 4$
(c) $X = 4$ or $X = 11$
(d) $X \leqslant 4$ or $X \geqslant 11$

A large company is considering introducing a new selection procedure for job applicants. The selection procedure is intended to result over a long period in equal numbers of men and women being offered jobs. The new procedure is tried with a random sample of applicants and 15 of them, 11 women and 4 men, are offered jobs.

(e) Carry out a suitable test at the 5% level of significance to determine whether it is reasonable to suppose that the selection procedure is performing as intended. You should state the null and alternative hypotheses under test and explain carefully how you arrive at your conclusions.

(f) Suppose now that, of the 15 applicants offered jobs, w are women. Find all the values of w for which the selection procedure should be judged acceptable at the 5% level.

[MEI]

HYPOTHESIS TEST FOR THE MEAN OF A POISSON DISTRIBUTION

The parameter λ of a Poisson distribution may be tested in a similar way to the preceding tests for the probability p of a binomial distribution. As before, both one-tail and two-tail tests may be encountered.

EXAMPLE 5.4

Over a long period of time it has been established that the number of minor accidents per month at a certain traffic junction may be modelled by a Poisson distribution with parameter 5.0.

During a period of road maintenance the local council decides to introduce new safety features at the junction, and in the first month afterwards it is observed that exactly one minor accident occurs.

(a) State formal hypotheses which may be tested.
(b) Find the critical region for a 5% hypothesis test.
(c) Conduct the test, stating your conclusion clearly.
(d) What criticism might be raised about using data from the first month after the maintenance has been carried out?

Solution **(a)** Let the new model be a Poisson distribution with parameter λ.

$$H_0: \lambda = 5.0$$
$$H_1: \lambda < 5.0.$$

(b) The critical region can be found from cumulative Poisson tables.

$\lambda =$	0.05	0.10	0.15	0.20	0.25	0.30	0.35	0.40	0.45	0.50
$x = 0$	0.6065	0.3679	0.2231	0.1353	0.0821	0.0498	0.0302	0.0183	0.0111	0.0067
1	0.9098	0.7358	0.5578	0.4060	0.2873	0.1991	0.1359	0.0916	0.0611	(0.0404)
2	0.9856	0.9197	0.8088	0.6767	0.5438	0.4232	0.3208	0.2381	0.1736	0.1247
3	0.9982	0.9810	0.9344	0.8571	0.7576	0.6472	0.5366	0.4335	0.3423	0.2650
4	0.9998	0.9963	0.9814	0.9473	0.8912	0.8153	0.7254	0.6288	0.5321	0.4405

FIGURE 5.8

Since 0.0404 is the last value below 5%, the critical region ends at 1. Thus the critical region is {0, 1}.

(c) The test value $X = 1$ lies in the critical region. Therefore reject H_0 in favour of H_1. The mean number of minor accidents does seem to have reduced.

(d) The first month after the maintenance is not a random month and might not be representative. For example, drivers might be driving cautiously because of a new road layout. It would have been better to wait for some time, then choose a month at random.

EXAMPLE 5.5

Kevin is the manager of a local football team. He examines the number of goals scored per match over the previous ten years and finds that a good model is a Poisson distribution with parameter 1.5.

During pre-season training Kevin introduces some new methods and in the first match of the new season the team scores four goals. Does this provide significant evidence, at the 5% level, that the mean number of goals has now increased?

Solution Assume that a Poisson model with parameter λ can be used to model the distribution of goals scored after the new training methods have been introduced.

H_0: $\lambda = 1.5$
H_1: $\lambda > 1.5$.

The critical region can be found from cumulative Poisson tables.

$\lambda =$	0.5	1.0	1.5	2.0	2.5	3.0	3.5	4.0	4.5	5.0
$x = 0$	0.6065	0.3679	0.2231	0.1353	0.0821	0.0498	0.0302	0.0183	0.0111	0.0067
1	0.9098	0.7358	0.5578	0.4060	0.2873	0.1991	0.1359	0.0916	0.0611	0.0404
2	0.9856	0.9197	0.8088	0.6767	0.5438	0.4232	0.3208	0.2381	0.1736	0.1247
3	0.9982	0.9810	0.9344	0.8571	0.7576	0.6472	0.5366	0.4335	0.3423	0.2650
4	0.9998	0.9963	0.9814	0.9473	0.8912	0.8153	0.7254	0.6288	0.5321	0.4405
5	1.0000	0.9994	0.9955	0.9834	0.9580	0.9161	0.8576	0.7851	0.7029	0.6160
6	1.0000	0.9999	0.9991	0.9955	0.9858	0.9665	0.9347	0.8893	0.8311	0.7622
7	1.0000	1.0000	0.9998	0.9989	0.9958	0.9881	0.9733	0.9489	0.9134	0.8666

FIGURE 5.9

The first value above 95% is 0.9814 corresponding to $X = 4$; this marks the upper limit of the non-critical region. Thus the critical region is {5, 6, 7, . . .}.

The test value of $X = 4$ does not lie in the critical region, so you accept H_0. There is no evidence that the mean number of goals has increased.

EXAMPLE 5.6

The number of defective cups produced per day at a small china factory is Poisson distributed with mean 8.5. After servicing the equipment the manager suspects that this figure might have changed, so he chooses a day's production at random, and finds that five defective cups were produced.

Conduct a hypothesis test to see if there is any evidence that the mean number of defective cups per day has changed. Use a 2% significance level.

Solution Assume that a Poisson model with parameter λ can be used to model the distribution of defective cups after the equipment has been serviced.

$$H_0: \lambda = 8.5$$
$$H_1: \lambda \neq 8.5$$

The critical region can be found from cumulative Poisson tables.

λ =	5.5	6.0	6.5	7.0	7.5	8.0	8.5	9.0	9.5	10.0
x = 0	0.0041	0.0025	0.0015	0.0009	0.0006	0.0003	0.0002	0.0001	0.0001	0.0000
1	0.0266	0.0174	0.0113	0.0073	0.0047	0.0030	0.0019	0.0012	0.0008	0.0005
2	0.0884	0.0620	0.0430	0.0296	0.0203	0.0138	0.0093	0.0062	0.0042	0.0028
3	0.2017	0.1512	0.1118	0.0818	0.0591	0.0424	0.0301	0.0212	0.0149	0.0103
4	0.3575	0.2851	0.2237	0.1730	0.1321	0.0996	0.0744	0.0550	0.0403	0.0293
5	0.5289	0.4457	0.3690	0.3007	0.2414	0.1912	0.1496	0.1157	0.0885	0.0671
6	0.6860	0.6063	0.5265	0.4497	0.3782	0.3134	0.2562	0.2068	0.1649	0.1301
7	0.8095	0.7440	0.6728	0.5987	0.5246	0.4530	0.3856	0.3239	0.2687	0.2202
8	0.8944	0.8472	0.7916	0.7291	0.6620	0.5925	0.5231	0.4557	0.3918	0.3328
9	0.9462	0.9161	0.8774	0.8305	0.7764	0.7166	0.6530	0.5874	0.5218	0.4579
10	0.9747	0.9574	0.9332	0.9015	0.8622	0.8159	0.7634	0.7060	0.6453	0.5830
11	0.9890	0.9799	0.9661	0.9467	0.9208	0.8881	0.8487	0.8030	0.7520	0.6968
12	0.9955	0.9912	0.9840	0.9730	0.9573	0.9362	0.9091	0.8758	0.8364	0.7916
13	0.9983	0.9964	0.9929	0.9872	0.9784	0.9658	0.9486	0.9261	0.8981	0.8645
14	0.9994	0.9986	0.9970	0.9943	0.9897	0.9827	0.9726	0.9585	0.9400	0.9165
15	0.9998	0.9995	0.9988	0.9976	0.9954	0.9918	0.9862	0.9780	0.9665	0.9513
16	0.9999	0.9998	0.9996	0.9990	0.9980	0.9963	0.9934	0.9889	0.9823	0.9730
17	1.0000	0.9999	0.9998	0.9996	0.9992	0.9984	0.9970	0.9947	0.9911	0.9857
18	1.0000	1.0000	0.9999	0.9999	0.9997	0.9993	0.9987	0.9976	0.9957	0.9928
19	1.0000	1.0000	1.0000	1.0000	0.9999	0.9997	0.9995	0.9989	0.9980	0.9965
20	1.0000	1.0000	1.0000	1.0000	1.0000	0.9999	0.9998	0.9996	0.9991	0.9984
21	1.0000	1.0000	1.0000	1.0000	1.0000	1.0000	0.9999	0.9998	0.9996	0.9993
22	1.0000	1.0000	1.0000	1.0000	1.0000	1.0000	1.0000	0.9999	0.9999	0.9997

FIGURE 5.10

To conduct a 2% two-tail test you look for 1% in each tail. At the lower end the critical value is 2, since 0.0093 is the last figure below 0.01. At the upper end the value 0.9934 corresponding to X = 16 is the first over 0.99, so this marks the end of the non-critical region. The upper critical region therefore begins at 17.

The full critical region is thus $\{0, 1, 2, 17, 18, 19, \ldots\}$.

Since $X = 5$ is not in the critical region you accept H_0.

There is no evidence that the mean has changed.

Exercise 5D

1 A gardener has divided a large lawn into squares 1 m by 1 m. The number of earthworms found in each square metre is Poisson distributed with mean 9.5.

(a) Find the probability that a randomly chosen square contains

(i) exactly six earthworms

(ii) no more than six earthworms.

(b) After a period of dry weather the gardener suspects that the mean number of earthworms might have decreased. He selects a square at random and finds that it contains exactly three earthworms.

(i) State suitable null and alternative hypotheses which the gardener could test.

(ii) Conduct the test at the 5% significance level, stating your conclusions clearly.

2 The number of cars which break down per day on a certain stretch of motorway is Poisson distributed with mean 5.5.

(a) Find the probability that, on a randomly chosen day,

(i) exactly five cars break down

(ii) less than three cars break down.

(b) During a particularly busy holiday period a traffic police officer reckons the mean is likely to have increased. To test this idea, he chooses a day at random, during the busy period, and finds that exactly nine cars break down.

(i) State suitable null and alternative hypotheses which the police officer could test.

(ii) Conduct the test at the 5% significance level, stating your conclusions clearly.

3 An orbiting space telescope is designed to discover Gamma Ray Bursters, or GRBs. During the two years of its operational lifetime the number of discoveries per month appears to follow a Poisson distribution with mean 1.5.

(a) Find the probability that, in a randomly chosen month, the telescope discovers

(i) exactly two GRBs

(ii) at least two GRBs.

At the end of the two years the telescope is serviced by a crew of astronauts who replace various components in order to extend its lifetime. During the following month it discovers five GRBs.

(b) Conduct a hypothesis test to see if there is any evidence that the mean number of discoveries per month has increased as a result of the service. Use a 5% significance level.

4 A fisherman reckons that the number of fish he catches per day can be modelled by a Poisson distribution with parameter 2.5.

(a) Assuming this model to be valid, calculate the probability that he catches at least three fish on a randomly chosen day.

After changing his bait the fisherman goes out for a day and catches k fish. He concludes that the mean number of fish he catches per day has increased as a result of the new bait, using a 5% significance level.

(b) Find the possible values of k.

5 A small internet website devoted to cookery receives, on average, 6.5 hits per day, over a long period of time.

(a) Find the probability that, on a randomly chosen day, the website receives at least five hits.

The site then receives some extra publicity by being featured in a cookery magazine. The webmaster suspects that this will result in an increase in the daily number of hits. He chooses a day at random and finds that the site receives 15 hits.

(b) State null and alternative hypotheses, explaining why your alternative hypothesis takes the form it does.

(c) Conduct the test at the 1% significance level. State your conclusion clearly.

EXERCISE 5E Examination-style questions

1 Explain briefly what you understand by

(a) a statistical model

(b) a sampling unit

(c) a sampling frame. [Edexcel]

2 The committee of a squash club are discussing whether or not to incorporate a sauna as part of the club's facilities. It is decided to ask a sample of members for their views.

(a) Suggest a suitable sampling frame.

(b) Explain the difference between a sample and a census, using this example to illustrate your answer.

(c) State why you think the committee might have decided to take a sample in this case rather than a census. [Edexcel]

3 A school held a disco for Years 9, 10 and 11 which was attended by 500 pupils. The pupils were registered as they entered the disco. The disco organisers were keen to assess the success of the event. They designed a questionnaire to obtain information from those who attended.

(a) State one advantage and one disadvantage of using a sample survey rather than a census.

(b) Suggest a suitable sampling frame.

(c) Identify the sampling units. [Edexcel]

4 Amy is a student at a sixth form college. She wants to conduct a survey to investigate the extent to which the college students smoke. The table below gives some information about the number of students at the college.

	Boys	Girls	Total
Lower sixth (Year 12)	180	160	340
Upper sixth (Year 13)	144	116	260
Total	324	276	600

Amy wants to sample 60 students in total.

(a) Explain why Amy should not sample 30 boys and 30 girls chosen at random.

(b) Draw up a table to show how many boys and girls from each year should be included in Amy's sample.

(c) Suggest a suitable sampling frame.

5 An optical firm produces lenses to be used in high quality microscope eyepieces. Ten per cent of the lenses produced contain manufacturing flaws and have to be discarded.

(a) Find the probability that, in a sample of 20, exactly four have to be discarded.

Following a change in the manufacturing process, the firm believes that the percentage of flawed lenses has been reduced. To test this hypothesis, a random sample of 50 lenses is taken. Three lenses are found to be flawed.

(b) Conduct a hypothesis test, at the 5% significance level, to see if there is any evidence that the percentage of flawed lenses has been reduced.

6 A newspaper carries a story claiming that 45% of the voting population support the opposition party. Peter suspects that this figure has been made up, so to test the validity of this claim he asks a random sample of 50 voters who they support. 32 of them say that they support the opposition party.

(a) Write down suitable null and alternative hypotheses.

(b) Conduct the test using a 5% significance level. Indicate clearly whether you are performing a one-tail or two-tail test.

7 On a certain motorway some of the cat's eyes (reflectors in the road) are missing. On average, there are six missing per kilometre of motorway.

(a) Find the probability that

(i) there are at least four missing from a given kilometre of motorway

(ii) there are exactly ten missing from a given 2 km stretch of motorway.

An engineer wants to see if a second motorway has a similar average number of missing cat's eyes. He selects a 1 km stretch of this second motorway and finds that there are 11 missing cat's eyes.

(b) Write down suitable null and alternative hypotheses.

(c) Conduct the test using a 5% significance level.

8 A cereal manufacturer puts a toy racing car in each box. The cars are either red, green, blue or yellow. Alicia thinks that yellow cars are found more than 25% of the time, so she collects 20 cereal boxes at random and finds that there are yellow cars inside nine of them.

(a) Write down suitable null and alternative hypotheses that Alicia could test.

(b) Conduct the test using a 2.5% significance level.

(c) State your conclusion plainly.

9 A book contains, on average, one misprint every ten pages.

(a) Find the probability that a randomly chosen page contains no misprints.

(b) Find the probability that ten randomly chosen pages contain no misprints.

A librarian wishes to see whether a second book contains a similar rate of occurrence of misprints. She looks at 30 randomly chosen pages from this second book, and finds that there are six misprints.

Let λ denote the average number of misprints, per 30 pages, in the second book.

(c) Write down null and alternative hypotheses which could be tested.

(d) Carry out the test at the 10% significance level.

10 In Manuel's restaurant the probability of a customer asking for a vegetarian meal is 0.30. During one particular day in a random sample of 20 customers at the restaurant three ordered a vegetarian meal.

(a) Stating your hypotheses clearly, test, at the 5% level of significance, whether or not the proportion of vegetarian meals ordered that day is unusually low.

Manuel's chef believes that the probability of a customer ordering a vegetarian meal is 0.10. The chef proposes to take a random sample of 100 customers to test whether or not there is evidence that the proportion of vegetarian meals ordered is different from 0.10.

(b) Stating your hypotheses clearly, use a suitable approximation to find the critical region for this test. The probability for each tail of the region should be as close as possible to 2.5%.

(c) State the significance level of this test giving your answer to 2 significant figures.

[Edexcel]

KEY POINTS

1 There are essentially two reasons why you might wish to take a sample:

- to estimate the values of the parameters of the parent population
- to conduct a hypothesis test.

2 When taking a sample you should ensure that:

- the data are relevant
- the data are unbiased
- the data are not distorted by the act of collection
- a suitable person is collecting the data
- the sample is of a suitable size
- a suitable sampling procedure is being followed.

3 Some sampling procedures are:

- simple random sampling
- stratified sampling
- cluster sampling
- systematic sampling
- quota sampling.

4 Other important definitions include:

- sampling frame – a listing of all the items available for sampling
- sampling units – the individuals to be sampled
- sampling fraction – the proportion of the population which is sampled.

5 Steps for conducting a hypothesis test:

- establish the null and alternative hypotheses
- check whether a one-tail or two-tail test is required
- identify the critical region
- check whether the data collected falls outside or inside the critical region
- accept or reject H_0
- interpret the result in plain English.

ANSWERS

CHAPTER 1

EXERCISE 1A (Page 5)

1 0.234
2 0.271
3 0.214
4 0.294
5 0.146
6 (a) 0.125
(b) 0.375
(c) 0.375
(d) 0.125
7 (a) 0.2461
(b) Exactly 7 (0.117 compared with 0.055)
8 (a) (i) 0.058
(ii) 0.198
(iii) 0.296
(iv) 0.448
(b) 2
9 (a) (i) 0.264
(ii) 0.368
(iii) 0.239
(iv) 0.129
(b) Assumed the probability of being born in January = $\frac{31}{365}$. This ignores the possibility of leap years and seasonal variations in the pattern of births throughout the year.
10 The three possible outcomes are not equally likely: 'one head and one tail' can arise in two ways (HT or TH) and is therefore twice as probable as 'two heads' (or 'two tails').

EXERCISE 1B (Page 6)

1 (a) $X \sim B(36, \frac{1}{6})$
(b) 6
(c) 5; 2.236
2 (a) 0.180
(b) 4
(c) 3
3 (a) 1.6
(b) 1.28
4 (a) 0.098
(b) 5.4
(c) 0.54
5 (a) There are a fixed number of independent trials with a constant probability of success/failure.
(b) 3.2
(c) 2.944
(d) 0.116
6 0.75
7 50
8 0.8 or 0.2

EXERCISE 1C (Page 11)

1 (a) 0.297
(b) 0.922
2 (a) 0.021
(b) 0.025
3 (a) 0.075
(b) 11; 4.95
(c) 0.409
4 (a) 0.060
(b) 0.258
(c) 0.498
5 (a) 0.233
(b) 0.194
(c) 0.930
6 (a) There is a fixed number of independent trials with a constant probability of success/failure. $X \sim B(5, 0.5)$
(b) 0.313
(c) 0.813
7 (a) 0.169
(b) 0.102
(c) 0.225
8 (a) 0.240
(b) 0.412
(c) 0.265
(d) 0.512

(e) 0.384
(f) 0.096
(g) 0.317
Assume that men/women in the office represent a random sample, as far as their weight is concerned.
9 (a) 0.003
(b) 0.121
(c) 0.167
(d) 0.723
10 (a) (i) 0.349
(ii) 0.387
(iii) 0.194
(b) 0.070
(c) 0.678

Exercise 1D (Page 12)

1 (a) (i) 0.000 13
(ii) 0.0322
(iii) 0.402
(b) 0 or 1 equally likely
2 (a) 2
(b) 0.388
(c) 0.323
3 (a) (i) 0.0123
(ii) 0.0988
(iii) 0.296
(iv) 0.395
(b) 2 min 40 s
4 (a) He must be an even number of steps from the bar. (The numbers of steps he goes east or west are either both even or both odd, since their sum is 12, and in both cases the difference between them, which gives his distance from the bar, is even.)
(b) 0.002 93
(c) At the bar
(d) 0.242
5 (a) 0.0735
(b) 2.7
(c) 0.267
6 (a) 0.010
(b) 0.317
(c) 0.346
(d) 0.922
7 (a) (i) 0.195
(ii) 0.299
(b) 3; 0.260
8 (a) $\frac{1}{64} = 0.0156$
(b) $\frac{15}{64} = 0.234$
(c) $\frac{45}{512} = 0.088$
(d) $\frac{45}{2048} = 0.022$
(e) $\frac{405}{8192} = 0.049$
9 (a) Binomial, $n = 20$, $p = 0.2$
(b) 0.0115
(c) 0.0321
10 (a) Binomial, $n = 10$, $p = 0.35$
(b) 0.262
(c) 0.0949
(d) The value of p will need to be increased.

Chapter 2

Exercise 2A (Page 22)

1 Using Po(4), 0.195
2 Using Po(2.4), 0.261
3 (a) 0.140
(b) 0.175
(c) 0.175
4 (a) 0.238
(b) 0.222
(c) 0.156
5 0.0536

Exercise 2B (Page 25)

1 (a) 0.050
(b) 0.149
(c) 0.224
(d) 0.423
(e) 0.577
2 (a) 0.342
(b) 0.658
3 X may be modelled by a Poisson distribution when cars arrive singly and independently and at a known overall average rate.
0.109
4 0.790
It is assumed that calls arrive singly and independently and with a known overall average rate of 4.2 calls per night.
5 (a) 42
(b) 2, 2.381
(c) 0.135, 0.271, 0.271, 0.180, 0.090, 0.036, 0.012
(d) 5.7, 11.4, 11.4, 7.6, 3.8, 1.5, 0.5
(e) Yes, because there is a good fit between the actual data and the predictions made in part (d).
6 (a) 0.111, 0.244, 0.268, 0.377
(b) 3

7 (a) 0.738
(b) 0.478, 0.353, 0.130, 0.032, 0.006, 0.001
(c) 239.0, 176.4, 65.1, 16.0, 3.0, 0.4
(d) Yes, there seems to be reasonable agreement between the actual data and the Poisson predictions.

8 (a) 0.333
(b) 0.002

9 600 m;
Po(2.5); 0.082, 0.109; 0.779, 0.207

10 Injuries are assumed to occur singly, independently and randomly.
0.5, 0.48; because the mean is approximately equal to the variance.
31.5, 15.8, 3.9, 0.7, 0.1

EXERCISE 2C (Page 30)

1 (a) 0.362
(b) 0.544
(c) 0.214
(d) 0.558

2 (a) 0.082
(b) 0.891
(c) 0.287

3 (a) 0.161
(b) 0.554
(c) 10
(d) 0.016
(e) 0.119

4 (a) 0.102
(b) 0.285
(c) 0.696

5 (a) 0.175
(b) 0.973
(c) 0.031
0.125; 0.249

6 (a) 0.175
(b) 0.560
(c) 0.125
(d) 0.542
(e) 0.0308
10

7 Some bottles will contain two or more hard particles. This will decrease the percentage of bottles that have to be discarded.
13.9%
Assume the hard particles occur singly, independently and randomly.

8 $X \sim \text{Po}(1)$; 0.014; 0.205

9 (a) 0.135
(b) 0.947
(c) 0.0527
(d) 13
(e) 0.876

10 (a) (i) 0.082
(ii) 0.456
(b) 0.309
(c) 1 east-bound and 2 west-bound

EXERCISE 2D (Page 33)

1 (a) 3
(b) 27.5
(c) 461

2 (a) It is assumed that incidents of criminal damage occur singly, independently and at a constant mean rate.
(b) (i) 0.271
(ii) 0.018
(iii) 0.184

3 (a) 0.472
(b) 0.041

4 (a) (i) The distribution of X is approximately binomial, n is large, p is small and $\lambda = np = 20$.
(ii) 0.857
(b) 231

5 7.03, 6.84;
The mean is approximately equal to the variance.
8.17

6 (a) 0.223
(b) 4
(c) 0.0498
(d) 6

7 (a) The mean is much greater than the variance therefore X does not have a Poisson distribution.
(b) Yes because now the values of the mean and variance are similar.
(c) 0.012

8 (a) X can be modelled by a Poisson distribution because the passage of cars happens independently, at random and at a constant mean rate. Also the values of the mean and variance are similar.
(b) 0.224
(c) 0.586

9 (a) 0.27
(b) 0.350
(c) 0.182

(d) 0.124; demand can be met provided not more than three cars are requested on any one day.

(e) £32.42

10 3.87, 3.53, taking $f_9 = 8$

Poisson distribution with parameter 3.87 because radioactive atoms decay randomly and independently and at a constant mean rate if the half-life is long compared with the duration of the experiment.

10.8, 41.9, 81.1, 104.6, 101.2, 78.3, 50.5, 27.9. 13.5, 9.2

There is quite good agreement between the two sets of figures.

75.8

CHAPTER 3

EXERCISE 3A (Page 44)

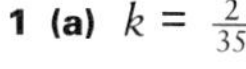

1 (a) $k = \frac{2}{35}$

(b)

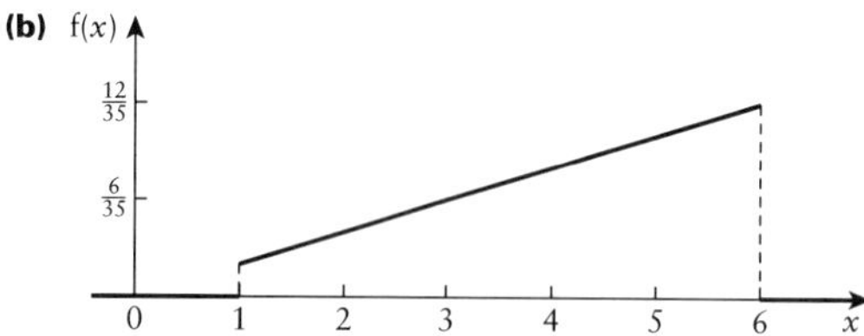

(c) $\frac{11}{35}$

(d) $\frac{1}{7}$

2 (a) $k = \frac{1}{12}$

(b)

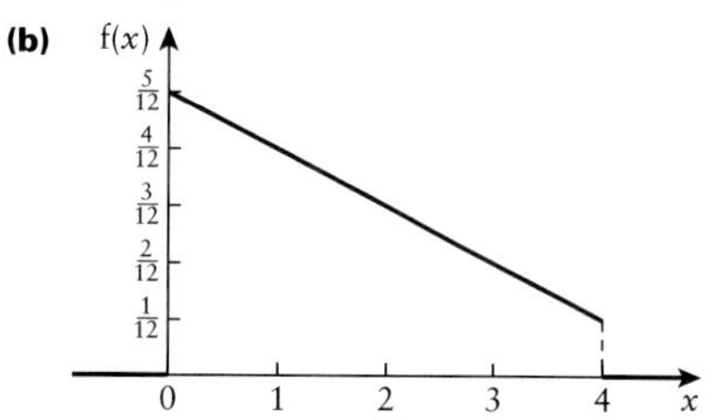

(c) 0.207

3 (a) $a = \frac{4}{81}$

(b)

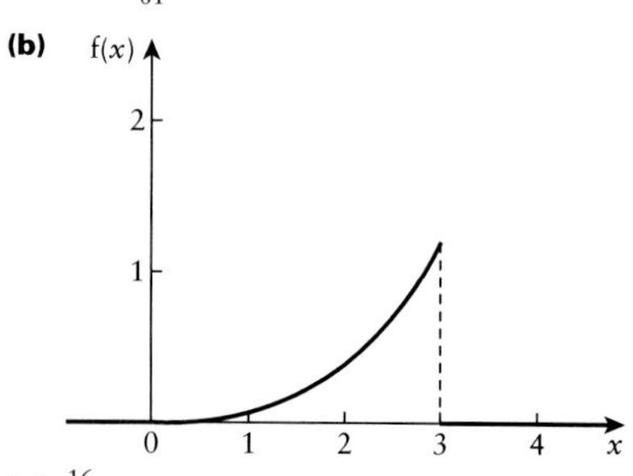

(c) $\frac{16}{81}$

4 (a) $k = \frac{1}{4}$

(b)

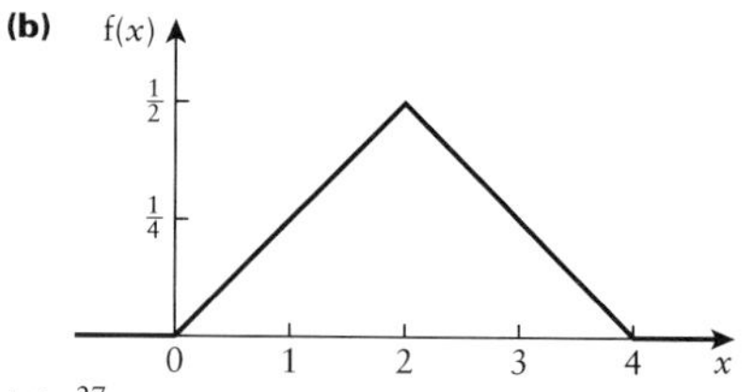

(c) $\frac{27}{32}$

5 (a) $c = \frac{1}{8}$

(b)

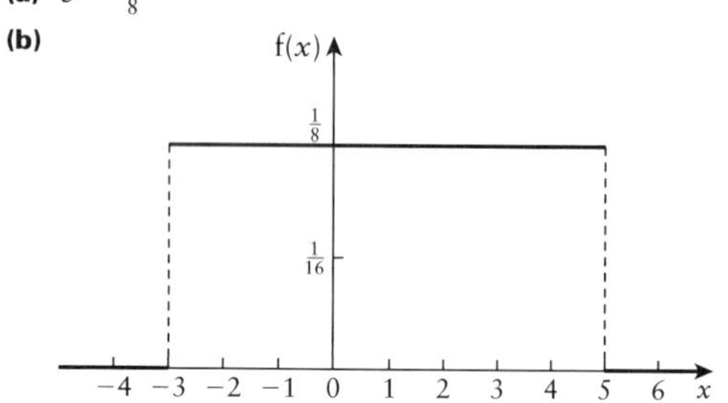

(c) $\frac{1}{4}$

(d) $\frac{3}{8}$

6 (a) $k = 0.048$

(b)

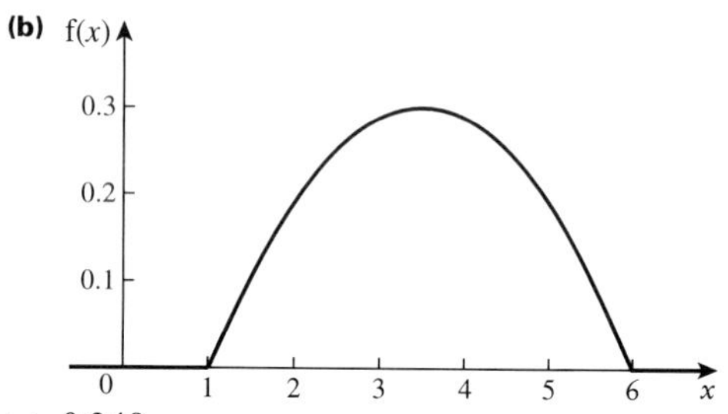

(c) 0.248

7 (a) $a = \frac{5}{12}$

(b)

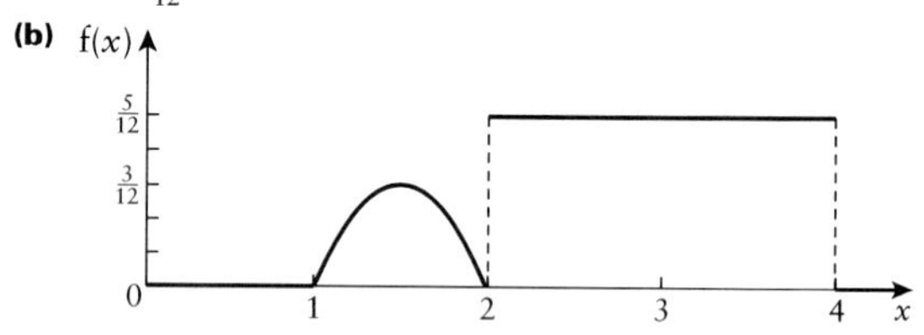

(c) 0.292

(d) $\frac{7}{12}$

8 (a) $k = \frac{2}{9}$

(b) 0.067

9 (a) $k = \frac{1}{100}$

(b)

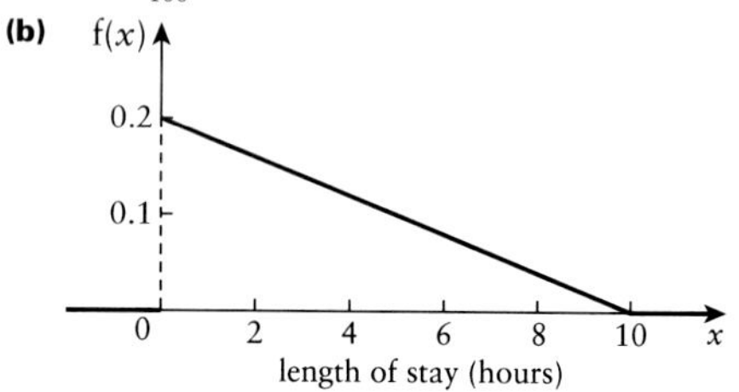

(c) 19, 17, 28, 36

(d) Yes

(e) Further information needed about the group 4–10 hours. It is possible that many of these stay all day and so are part of a different distribution.

10 (a)

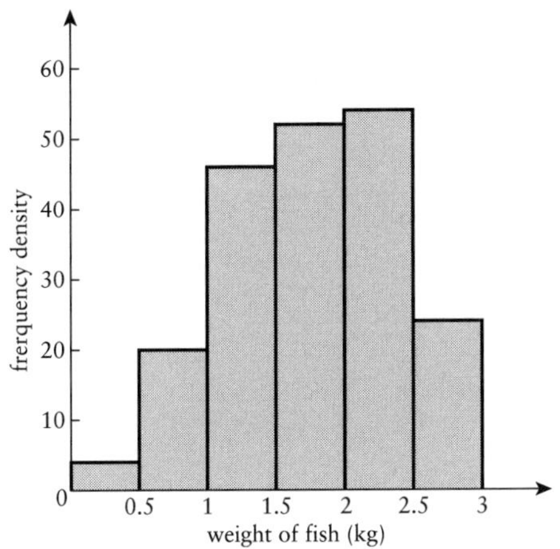

Negative skew

(b)

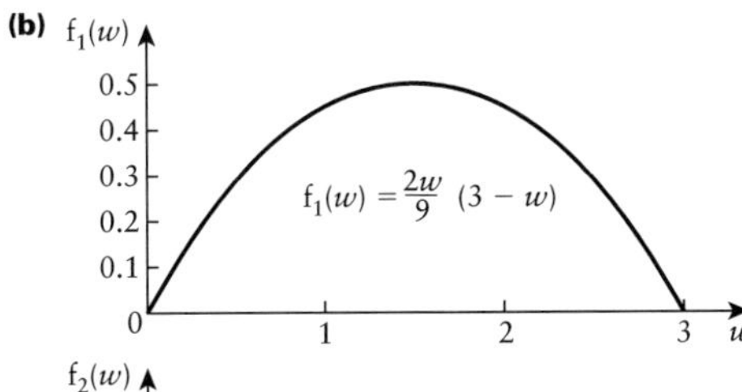

$f_2(w) = \frac{10w^2}{81}(3 - w)^2$

$f_3(w) = \frac{4w^2}{27}(3 - w)$

$f_4(w) = \frac{4w}{27}(3 - w)^2$

f_3

(c) 1.62, 9.49, 20.14, 28.01, 27.55, 13.19

(d) Model seems good.

11 (a) $a = 100$

(b) 0.045

(c) 0.36

12 0.803, 0.456

13 (a) 0, 0.1, 0.21, 0.12, 0.05, 0.02, 0

(b) **(i)** 0.1 **(ii)** 0.31 **(iii)** 0.33 **(iv)** 0.17 **(v)** 0.07 **(vi)** 0.02

(c) $k = \frac{1}{1728}$

(d) **(i)** 0.132 **(ii)** 0.275 **(iii)** 0.280 **(iv)** 0.201 **(v)** 0.095 **(vi)** 0.016

(e) Model quite good. Both positively skewed.

EXERCISE 3B (Page 56)

1 (a) 2.67

(b) 0.89

(c) 2.828

2 (a) 2

(b) 2

(c) 1.76

3 (a) 0.6

(b) 0.04

(c) $\frac{2}{3}$

4 (a)

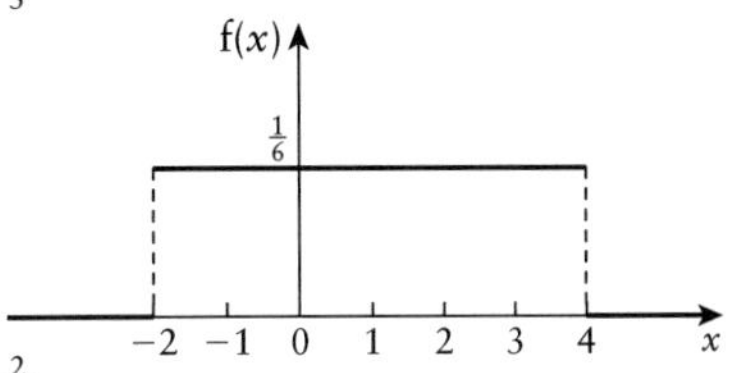

(b) $\frac{2}{3}$

(c) 1

(d) $\frac{1}{3}$

5 (a) 1.5

(b) 0.45

(c) 1.5

(d) 1.5

(e)

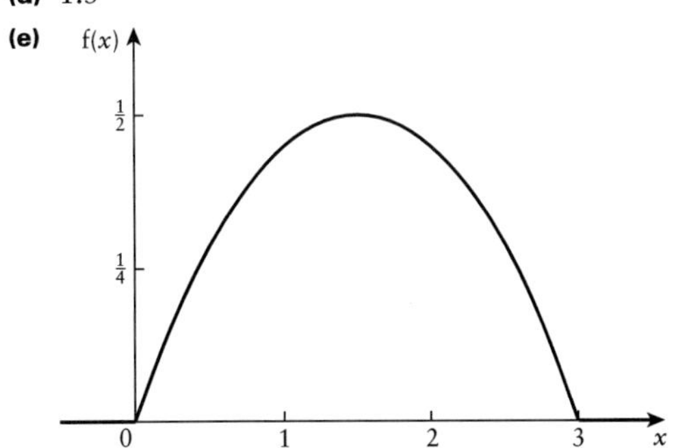

The graph is symmetrical and peaks when $x = 1.5$ thus $E(X)$ = mode of X = median value of $X = 1.5$.

6 (b) 1.083, 0.326

(c) 0.5625

7 (a) $f(x) = \frac{1}{10}$ for $10 \leqslant x \leqslant 20$

(b) 15, 8.33

(c) (i) 57.7%

(ii) 100%

8 (a)

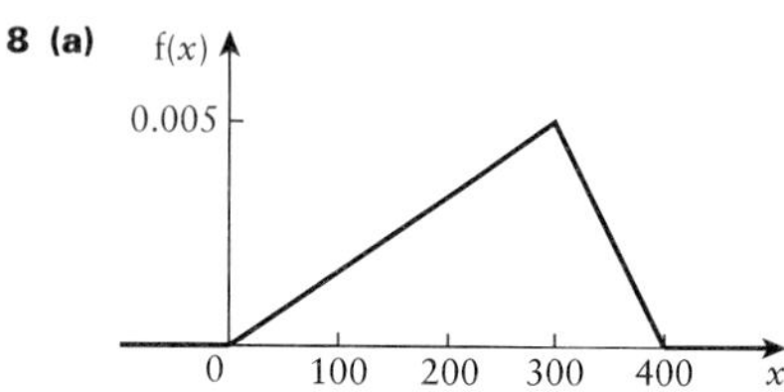

(c) 233 hours

(d) 7222.2

(e) 0.083

9 (a) $k = 1.2 \times 10^{-8}$

(b)

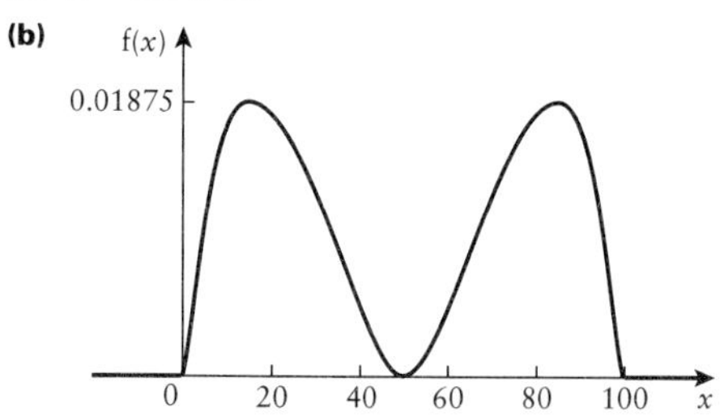

(c) The distribution is the sum of two smaller distributions, one of moderate candidates and the other of able ones.

(d) Yes if the step size is small compared to the standard deviation.

10 (b) 8.88, 2.88; 0.724

(c) $2m^3 - 18m^2 + 78m - 900 = 0$

11 (a)

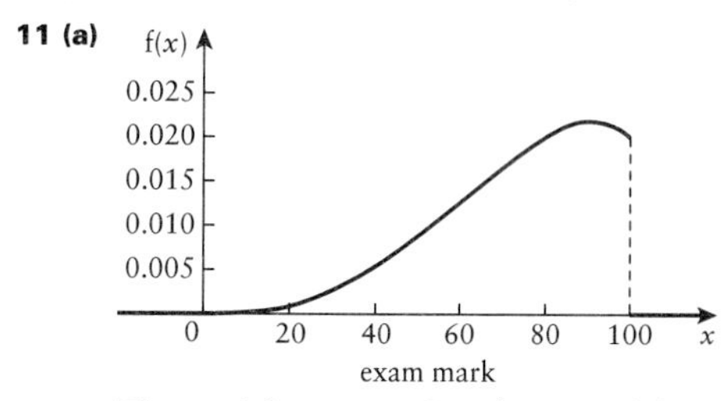

The model suggests that these candidates were generally of high ability as a large proportion of them scored a high mark.

(c) 12.5%

(d) No; 91

EXERCISE 3C (Page 68)

1 (a) 2.5

(b) $F(x) = 0$ for $x < 0$

$\frac{x}{5}$ for $0 \leqslant x \leqslant 5$

1 for $x > 5$

(c) 0.4

2 (a) $k = \frac{2}{39}$

(b)

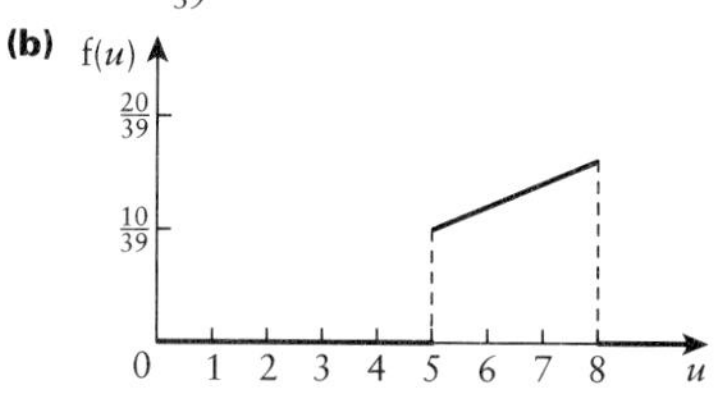

(c)
$$F(u) = 0 \quad \text{for } u < 5$$
$$= \frac{u^2}{39} - \frac{25}{39} \quad \text{for } 5 \leqslant u \leqslant 8$$
$$= 1 \quad \text{for } u > 8$$

(d)

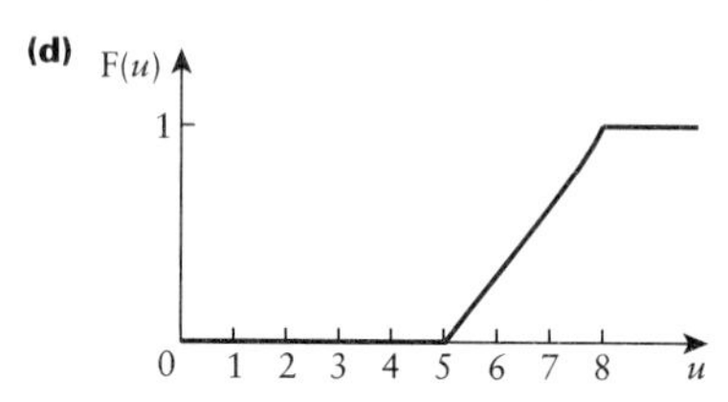

3 (a) $c = \frac{1}{21}$

(b)
$$F(x) = 0 \quad \text{for } x < 1$$
$$= \frac{x^3}{63} - \frac{1}{63} \quad \text{for } 1 \leqslant x \leqslant 4$$
$$= 1 \quad \text{for } x > 4$$

(c) 3.19

(d) 4

4 (a)
$$F(x) = 0 \quad \text{for } x < 0$$
$$= 1 - \frac{1}{(1+x)^3} \quad \text{for } x \geqslant 0;$$

(b) $x = 1$

5 (a) $\frac{1}{4}$

(b) 0.134

(c) $f(x) = 2 - 2x$ for $0 \leqslant x \leqslant 1$

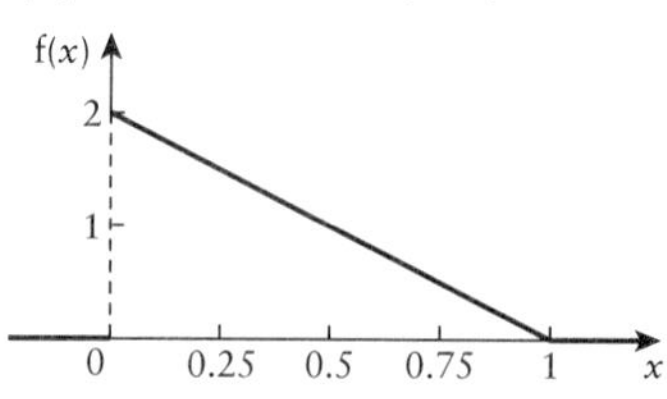

6 $E(X) = \frac{3}{4}$, $Var(X) = \frac{19}{80}$
$$F(x) = 0 \quad \text{for } x < 0$$
$$= \frac{3x}{4} - \frac{x^3}{16} \quad \text{for } 0 \leqslant x \leqslant 2$$
$$= 1 \quad \text{for } x > 2$$

7 $\frac{3}{5}$, 0.683

8 (b)
$$F(t) = 0 \quad \text{for } t < 0$$
$$= \frac{t^3}{432} - \frac{t^4}{6912} \quad \text{for } 0 \leqslant t \leqslant 12$$
$$= 1 \quad \text{for } t > 12$$

(d) 0.132

9 (a) 2.93

(b) $F(x) = 1 - \dfrac{(x-10)^2}{100}$ for $0 \leqslant x \leqslant 10$

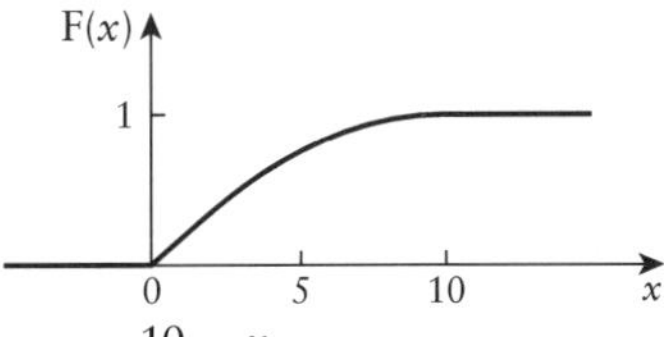

(c) $f(x) = \dfrac{10-x}{50}$ for $0 \leqslant x \leqslant 10$

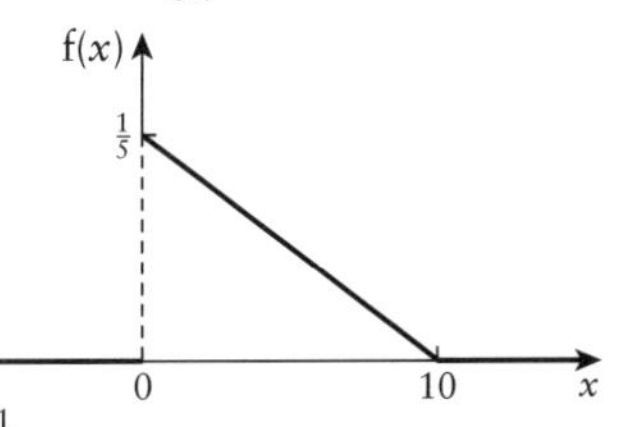

10 (a) $\frac{1}{3}$

(c) 4.39

(d) 12.5

EXERCISE 3D (Page 70)

1 (a) $f(x) = \dfrac{1}{2a}$ for $-a \leqslant x \leqslant a$

$= 0$ otherwise

(b) $0, \dfrac{a^2}{3}$

2 (a) $E(X) = 3.2$

(b)

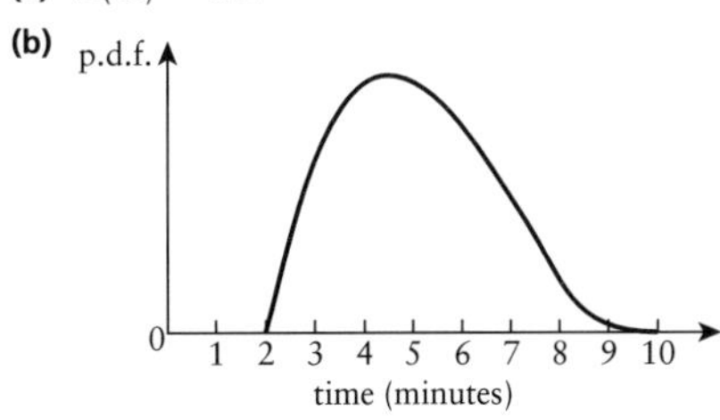

The model implies that all of the doctor's appointments last between 2 and 10 minutes, the mean time being 5.2 minutes and the variance of the distribution being 2.56 minutes^2.

3 $F(x) = \int f(x)\,dx$ and so $F'(x) = f(x)$.

Uniform distribution with mean value $\dfrac{a}{2}$.

$F(x) = 1 - \dfrac{(a-x)^2}{a^2}$ for $0 \leqslant x \leqslant a$

Mean of sum of smaller parts $= a$

4 (a) $\frac{5}{16}$

(b) $F(1.22) = 0.4925\ldots < 0.5$, whereas $F(1.23) = 0.5012 > 0.5$

(c) $f(t) = \begin{cases} \dfrac{3t^2}{2} - \dfrac{3t^3}{4} & 0 \leqslant t \leqslant 2 \\ 0 & \text{otherwise} \end{cases}$

(d) $\frac{4}{3}$

(e)

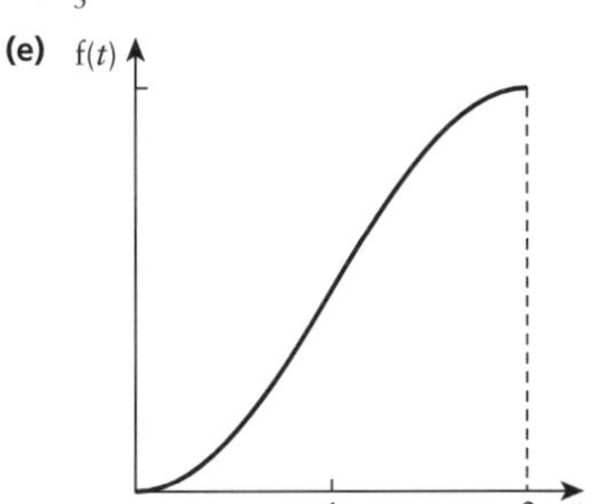

(f) On a windy day the taper is not likely to stay alight for so long

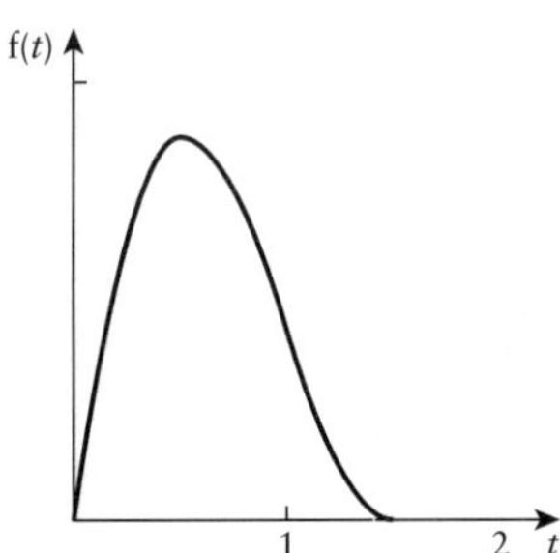

5 (a)

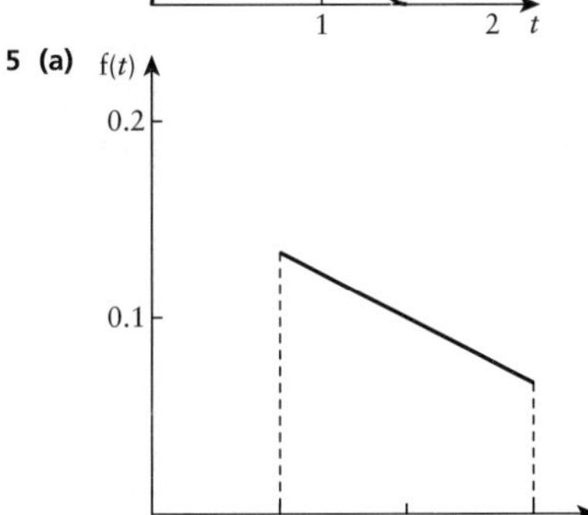

(b)

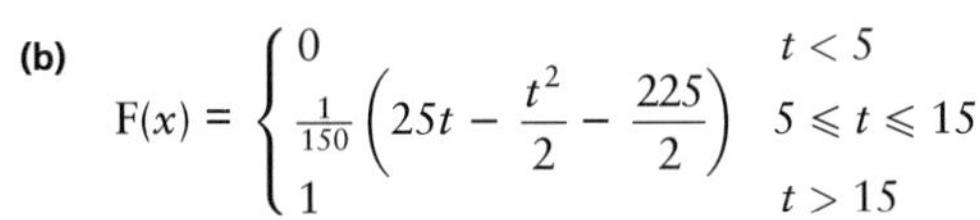

$$F(x) = \begin{cases} 0 & t < 5 \\ \frac{1}{150}\left(25t - \dfrac{t^2}{2} - \dfrac{225}{2}\right) & 5 \leqslant t \leqslant 15 \\ 1 & t > 15 \end{cases}$$

(c) 0.23

(e) The model may be unrealistic because it does not allow for very short or long calls. Perhaps a bimodal model would be better.

(f)

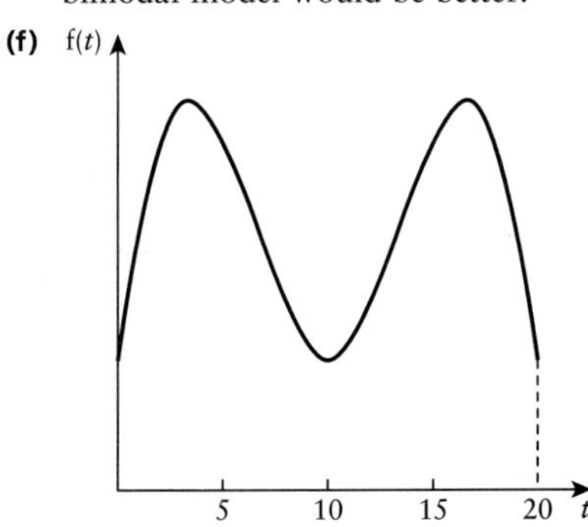

6 (a)

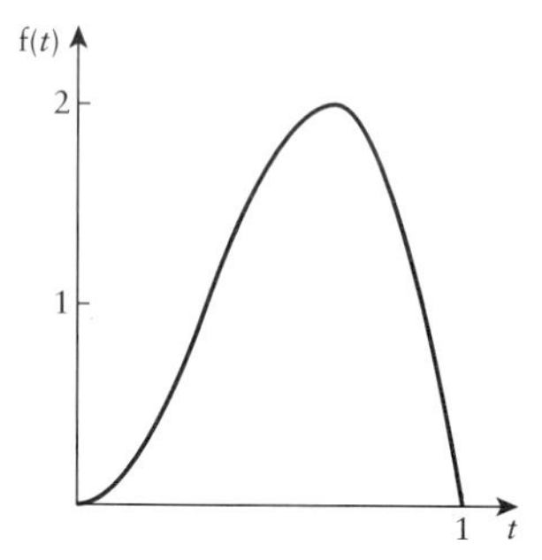

(b) The industrial estate is probably in an out-of-town location, since most of the workers take almost an hour to travel there.

(c) 40 minutes, 10.7 minutes

(d) $$\mathrm{F}(x) = \begin{cases} 0 & x < 0 \\ 5x^4 - 4x^5 & 0 \leqslant x \leqslant 1 \\ 1 & x > 1 \end{cases}$$

7 (a) 0.3125

(d) 0.0625

(e) 1.6

(f) The teacher is very unlikely to be under 1 m tall; the answer to part **(d)** is therefore more realistic than the answer to part **(a)**. The mean height of the teacher is more likely to be near 1.6 m than 1.25 m, so the answer to part **(e)** is therefore more realistic than the answer to part **(b)**.
Thus random variable X is likely to be the more appropriate.

8 (a) 3.8, 0.36

(b) 4

(c) 59.3%

9 (a) $\frac{2}{3}$, $\frac{1}{18}$

(b) $$\mathrm{F}(x) = \begin{cases} 0 & x < 0 \\ x^2 & 0 \leqslant x \leqslant 1 \\ 1 & x > 1 \end{cases}$$

(c) $1\frac{2}{3}$, $\frac{8}{9}$

(d) $-1 \leqslant x \leqslant 3$

(e) £1828

10 (a)

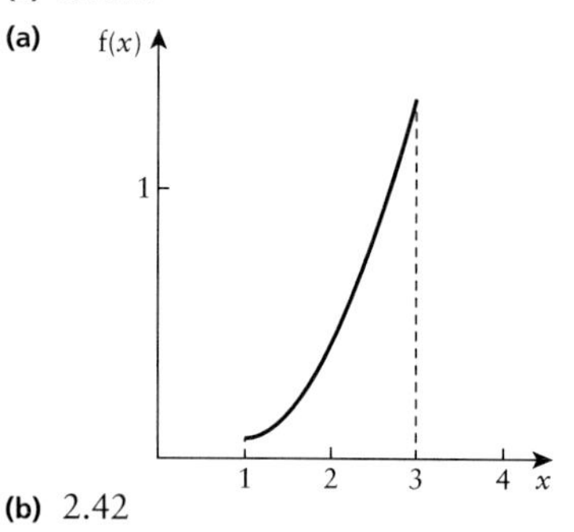

(b) 2.42

(d) $$\mathrm{F}(x) = \begin{cases} 0 & x < 1 \\ \frac{1}{80}(x^4 - 1) & 1 \leqslant x \leqslant 3 \\ 1 & x > 3 \end{cases}$$

(e) 0.654

(f) 0.612, which is in reasonably good agreement with the true value.

CHAPTER 4

EXERCISE 4A (Page 80)

1 (a) 10, 33.3

(b) $$\mathrm{f}(x) = \begin{cases} 0.05 & 0 \leqslant x \leqslant 20 \\ 0 & \text{otherwise} \end{cases}$$

(c) 0.35

2 (a) 15, 75

(b) $$\mathrm{f}(x) = \begin{cases} 0.0\dot{3} & 0 \leqslant x \leqslant 30 \\ 0 & \text{otherwise} \end{cases}$$

(c) $\frac{1}{3}$

3 (a) 0.6

(c) $$\mathrm{f}(x) = \begin{cases} 0.8\dot{3} & 0 \leqslant x \leqslant 1.2 \\ 0 & \text{otherwise} \end{cases}$$

(d) $\frac{1}{6}$

(e) 0.0625

4 (a) Random point implies a uniform distribution

(b) 25, 208.3

(c) $$\mathrm{f}(x) = \begin{cases} 0.02 & 0 \leqslant x \leqslant 50 \\ 0 & \text{otherwise} \end{cases}$$

(d) $$\mathrm{F}(x) = \begin{cases} 0 & x < 0 \\ 0.02x & 0 \leqslant x \leqslant 50 \\ 1 & x > 50 \end{cases}$$

(e) 0.2

5 (a) 0.4

(b) 0.2

EXERCISE 4B (Page 85)

1 99.4% of the population have IQs less than 2.5 standard deviations above the mean.
0.165

2 (a) 106

(b) 75 and 125

(c) 39.5

3 (a) 0.282

(b) 0.526

(c) (i) 25

(ii) 4.33

(d) 0.102

4 0.246, 0.0796, 0.0179
The normal distribution is used for continuous data; the binomial distribution is used for discrete data. If a normal approximation to the binomial distribution is used then a continuity correction must be made. Without this the result would not be accurate.

5 n must be large and p must not be too close to 0 or 1. These conditions ensure that the distribution is reasonably symmetrical so that its probability profile resembles a normal distribution.
(a) 0.1853
(b) 0.1838
(c) 0.81%

6 **(a)** **(i)** 0.315
(ii) 0.307; assuming the answer to part **(a) (i)** is correct, there is a 7.6% error; worse
(b) 0.5245

7 $\frac{1}{3}$; 6.667; 4.444; 13

8 **(a)** 0.590
(b) 0.081
It is assumed that the defective syringes are mixed randomly with the functional ones. Also, as the total number of syringes in the box is very large, removing one syringe leaves the probability distribution of defective and functional syringes unchanged.

It may be convenient to use the normal approximation as this simplifies the probability calculation. So long as a continuity correction is made the result ought to be very close to the binomial model result.
$P(X \geqslant 15) = 0.067$

EXERCISE 4C (Page 89)

1 $^{2000}C_N \left(\frac{1}{30}\right)^N \left(\frac{29}{30}\right)^{2000-N}$
86; more (96)

2 0.180; 60, 7.75, 0.9124

3 Mean $= \sum_{r=0}^{\infty} rP(X = r) = \lambda$, called the Poisson parameter.
Variance $= E(X^2) - \lambda^2 = \lambda$
The Poisson parameter should be greater than 10 so that the probability profile is approximately bell-shaped.
(a) 0.0222
(b) 0.9778

4 **(a)** X has a binomial distribution with $p = 0.8$ i.e. $X \sim B(n, 0.8)$; 0.9389
(b) $X \sim N(80, 16)$; 0.0845
(c) 71

5 **(a)** 2.5; assume that service calls occur singly, independently and randomly.
(b) 0.918, 0.358
(c) 0.158

6 **(a)** **(i)** 0.5134
(ii) 0.1443
(b) $Po(53.\dot{3})$ which is approximately $N(53.\dot{3}, 53.\dot{3})$
(c) 39 and 68

EXERCISE 4D (Page 91)

1 **(a)** 0.1047
(b) $\mu = 24, \sigma^2 = 14.4$
(c) 0.1048
(d) Very good agreement, to within about 0.1%

2 **(a)** 0.5438
(b) $Y \sim Po(150)$ which is approximately $N(150, 150)$
(c) 0.1956

3 **(a)** Uniform continuous distribution on [0, 12]
$$f(x) = \begin{cases} \frac{1}{12} & 0 \leqslant x \leqslant 12 \\ 0 & \text{otherwise} \end{cases}$$
(b)
$$F(x) = \begin{cases} 0 & x < 0 \\ \frac{1}{12}x & 0 \leqslant x \leqslant 12 \\ 1 & x > 12 \end{cases}$$
(c) $\frac{1}{3}$

4 **(a)** 3.35, 3.2775
(b) Accidents are liable to be random, independent, uniform. Also the mean and variance are approximately equal.
(c) 9.4 days
(d) 0.6594
(e) 0.0960
(f) Not abnormally high – this type of value is likely to occur about 10% of the time anyway, according to the answer to part **(e)**.

5 **(a)** 0.0353
(b) 49.35, 6.908
(c) 0.0533
(d) 38
(e) The probability of this happening is extremely low, so the local authority should be suspicious.

6 **(a)** 0.0563
(b) 0.2241
(c) 0.0919
(d) 0.0004 which is very low
(e) Bella's mean is 43.75 with a standard deviation of 3.75, so she is very likely to achieve a pass mark of 40 or more.

7 (a) The occurrences are random, independent and uniform.
(b) 0.1334
(c) 11
(d) 0.7119

8 (a) Fixed number of trials with a constant probability of success.
(b) (i) 0.1623
(ii) 0.2493
(c) 0.1452
(d) (i) Using Po(6), 0.0113
(ii) Using N(54, 29.7), 0.6790

9 (a) 20, 12.48
(b) 25, 14.43
(c) The mean and standard deviation of the model are both higher than for the data set, by about 25% and 16% respectively, so the model does not seem as particularly good one.
(d) f(x) has mean 16.67, standard deviation 11.79
g(x) has mean 33.33, standard deviation 11.79
thus f(x) is preferable because its mean is closer to that for the data set.

10 (a) Poisson distribution, with $\lambda = 2.5$
(b) 0.0821
(c) 0.1088
(d) The presence of the sheep dog means that the sheep are now unlikely to be uniformly (randomly) scattered throughout the field.
(e) 0.1094

CHAPTER 5

EXERCISE 5A (Page 101)

1 (a) Systematic sampling
(b) (i) Simple random sampling
(ii) $\frac{1}{25}$
(c) (i) Cluster sampling
(ii) No. The streets are chosen at random and then 15 houses are chosen at random. However, not every sample of size 15 (throughout the town) can be chosen.
(d) (i) Quota sampling
(ii) No
(iii) The sample is small. It is questionable how reliable such information would be.

2 (a) (i) Years 1 and 2: 7 students from each;
Years 3 and 4: 5 students from each;
Year 5: 6 students.
(ii) $\frac{1}{20}$
(b) (i) 28 light vans, 2 company cars and 1 large-load vehicle.
(ii) Randomly choose the appropriate number of vehicles from each type. This is stratified sampling.
(c) (i) $\frac{1}{8}$
(ii) 0–5: 5; 6–12: 10; 13–21: 13 or 14; 22–35: 25 or 26; 36–50: 22 or 23; 51+: 3 or 4.

3 (a) Cluster sampling. Choose representative streets or areas and sample from these streets or areas.
(b) Stratified sample. Identify routes of interest and randomly sample trains from each route.
(c) Stratified sample. Choose representative areas in the town and randomly sample from each area as appropriate.
(d) Stratified sample as in part (c).
(e) Depends on method of data collection. If survey is, say, via a postal enquiry, then a random sample may be selected from a register of addresses.
(f) Cluster sample. Routes and times are chosen and a traffic sampling station is established to randomly stop vehicles to test tyres.
(g) Cluster sampling. Areas are chosen and households are then randomly chosen.
(h) Cluster sampling. A period (or periods) is chosen to sample and speeds are surveyed.
(i) Cluster sampling. Meeting places for 18-year-olds are identified: night clubs, pubs etc. and samples of 18-year-olds are surveyed, probably via a method to maintain privacy. This might be a questionnaire to ascertain required information.
(j) Random sample. The school pupil list is used as a sampling frame to establish a random sample within the school.

4 (a) Systematic sampling. Easy to set up but may be difficult to track down the student once they have been identified.
(b) Stratified sampling. Will reflect all opinions, but only as defined by the surveyor. Easy to carry out. That is, it should be easy to access the desired sample, students as they enter the college premises.
(c) The sample will be biased. Easy to survey. Those using the canteen will be surveyed.
(d) Cluster sampling. Assumes first and second year students are representative of the whole college. (If there are only first and second year students this will be true. The sampling procedure is then stratified.) Similar to **(a)**, that is, once students have been chosen from the lists they have to be located to seek their views.

5 (a) All production lines are identified. If it is judged they are equivalent then one (or more) can be chosen to produce a sample. This is cluster sampling. From this (or these) production line(s) a day (or days) is chosen to be the time when a sample is taken. A reasonable number of strip lights is chosen and then tested to destruction, that is, tested until they are exhausted.
An estimate is found from the mean life of the sample chosen.
(b) The map of the forest is covered with a grid. Each grid square is numbered. A sample is chosen by randomly selecting the squares. The tree (or trees) in each of the chosen squares is sampled.
(c) (i) A sample of 100 chips to be taken from each production line each working day (assuming a five-day week).
(ii) Stratified.
(d) Depending on the number of staff, one could carry out a census of all staff or, if more appropriate, a stratified sample based on part-time staff, full-time staff, etc.
(e) Identify different courses in your school/college. Access pupils from each of the courses, choosing them at random in order to elicit their views. This is a stratified sample.

7 (a) 2; 1.183
(b) P(2 defectives in 10) = 0.302
In 50 samples of 10, the expected number of samples with two defectives is 15.1, which agrees well with the observed 15.
(c) H_0: P(mug defective) = 0.2
H_1: P(mug defective) < 0.2
$n = 20$. P(0 or 1 defective mug) = 0.011
Critical region is {0} Accept H_0
It is reasonable to suppose that the proportion of defective mugs has been reduced.
(d) Critical region is {0, 1} Reject H_0
8 (a) 0.430
(b) 0.9619
(c) 0.0046
(d) H_0: $p = 0.9$, H_1: $p < 0.9$
(e) Accept H_0. The service does not appear to have deteriorated.
(f) Critical region is {0, 1, ..., 10}, since $P(X \leqslant 10) = 0.0127$.
9 (a) (i) 0.0278
(ii) 0.0384
(b) Let p = P(blackbird is male)
H_0: $p = 0.5$, H_1: $p > 0.5$
(c) Reject H_0. Critical region is {12, 13, ..., 16}.
(d) You would be more reluctant to reject H_0. The sampling method is likely to give a non-random sample.
10 (a) (i) 0.0991
(ii) 0.1391
(b) Let p = P(seed germinates)
H_0: $p = 0.8$, H_1: $p > 0.8$, since a higher germination rate is suspected.
(c) Critical region is {17, 18}, since
$P(X \geqslant 17) = 0.0991 < 10\%$ but
$P(X \geqslant 16) = 0.2713 > 10\%$.
(d) (i) When $p = 0.8$ he reaches the wrong conclusion if he rejects H_0, i.e. if $X \geqslant 17$, with probability 0.0991.
(ii) When $p = 0.82$ he reaches the wrong conclusion if he fails to reject H_0, i.e. if $X \leqslant 16$, with probability $1 - 0.1391 = 0.8609$.

EXERCISE 5B (Page 108)

1 Critical region is {0, ..., 7} Accept H_0
2 (a) 0.180
(b) 0.416
H_0: probability of passing first test = 0.6
H_1: probability of passing first test < 0.6
$17 \leqslant N \leqslant 20$
3 Critical region is {9, 10} Accept H_0
4 H_0: probability that toast lands butter-side down = 0.5
H_1: probability that toast lands butter-side down > 0.5
Critical region is {11, ..., 15} Reject H_0. The toast does not seem more likely to land butter-side down.
5 Critical region is {0, ..., 10} Reject H_0
There is evidence that the complaints are justified at the 5% significance level, though Mr McTaggart might object that the candidates were not randomly chosen.
6 Critical region is {6, ..., 50} Accept H_0
Insufficient evidence at the 5% significance level that the machine needs servicing.

EXERCISE 5C (Page 113)

1 Critical region is {0, ..., 5} and {15, ..., 20} Accept H_0
2 Critical region is {0, ..., 3} and {12, ..., 15} Reject H_0
3 Critical region is {0, 1, 2} and {10, 11, 12} Accept H_0

4 5% critical region is $\{0, 1\}$ and $\{10, \ldots, 20\}$ Reject H_0 but data not independent.
5 Critical region is $\{0\}$ and $\{8, \ldots, 20\}$ Accept H_0, the complaint is not justified.
6 Critical region is empty at left tail so accept H_0
7 $\leqslant 1$ or >9 males
8 $\leqslant 1$ or >8 correct
9 Critical region is $\leqslant 3$ or $\geqslant 13$ letter Zs
10 (a) 20
(b) 0.0623
(c) 5% critical region is $\{0, \ldots, 15\}$ and $\{25\}$ Reject H_0
11 (a) 0.0417
(b) 0.0592
(c) 0.0833
(d) 0.1184
(e) Let p = P(man selected)
H_0: $p = 0.5$, H_1: $p \neq 0.5$
Critical region is $\{0, \ldots, 3\}$ and $\{12, \ldots, 15\}$
Accept H_0, it is reasonable to suppose that the process is satisfactory.
(f) $4 \leqslant w \leqslant 11$

Exercise 5D (Page 118)

1 (a) (i) 0.0764
(ii) 0.1649
(b) (i) Let new mean be λ.
H_0: $\lambda = 9.5$
H_1: $\lambda < 9.5$
(ii) Reject H_0, since critical region is $\{0, 1, 2, 3, 4\}$.
2 (a) (i) 0.1714
(ii) 0.0884
(b) (i) Let new mean be λ.
H_0: $\lambda = 5.5$
H_1: $\lambda > 5.5$
(ii) Accept H_0, since critical region is $\{11, 12, \ldots\}$.
3 (a) (i) 0.2510
(ii) 0.4422
(b) Let new mean be λ.
H_0: $\lambda = 1.5$
H_1: $\lambda > 1.5$
Reject H_0, since critical region is $\{5, 6, \ldots\}$.
4 (a) 0.4562
(b) Possible values of k are $\{6, 7, 8, \ldots\}$.
5 (a) 0.7763
(b) Let new mean be λ.
H_0: $\lambda = 6.5$
H_1: $\lambda > 6.5$
The webmaster suspects an increase in the mean.
(c) Reject H_0, since critical region is $\{14, 15, 16, \ldots\}$.

Exercise 5E (Page 119)

1 (a) A statistical model is a set of mathematical equations which are used to calculate the probabilities of various outcomes in a statistical experiment.
(b) A sampling unit is an individual surveyed in a sample survey.
(c) A sampling frame is a listing of all the units which could be included in a sample.
2 (a) the membership list
(b) A sample would comprise only some of the members on the list; a census would include them all.
(c) If there are a large number of members then a sample would be cheaper/easier/more efficient than working with the whole population.
3 (a) Advantage: sample survey is easier to collect/process. Disadvantage: can be prone to bias if not carried out carefully.
(b) the registration list
(c) The individual pupils from Years 9, 10, 11 who attended the disco.
4 (a) There are more boys than girls so she should not use equal numbers in her survey.
(b)

	Boys	Girls
Lower sixth (Year 12)	18	16
Upper sixth (Year 13)	14	12

(c) The college registration list/roll.
5 (a) 0.0898
(b) Accept H_0 since critical region is $\{0, 1\}$.
6 (a) Let p = probability that a voter supports the opposition party.
H_0: $p = 0.45$
H_1: $p \neq 0.45$
(b) Critical region is $\{0, 1, \ldots, 15\}$ and $\{30, 31, \ldots, 50\}$, i.e. a two-tail test.
Reject H_0. The figure of 45% does appear to have been made up.
7 (a) (i) 0.8488
(ii) 0.1048
(b) Let new mean be λ.
H_0: $\lambda = 6$
H_1: $\lambda \neq 6$
(c) Accept H_0, since critical region is $\{0, 1\}$ and $\{12, 13, 14, \ldots\}$.
8 (a) Let p = P(yellow car)
H_0: $p = 0.25$
H_1: $p > 0.25$

(b) Reject H_0, since critical region is $\{10, 11, \ldots, 20\}$.

(c) There is evidence, at the 2.5% significance level, to support Alicia's claim that yellow cars occur more than 25% of the time.

9 (a) 0.9048

(b) 0.3679

(c) Let new mean be λ.
H_0: $\lambda = 3$
H_1: $\lambda \neq 3$

(d) Accept H_0, since critical region is $\{0\}$ and $\{7, 8, 9, \ldots\}$.

10 (a) Let p = P(vegetarian meal)
H_0: $p = 0.30$
H_1: $p < 0.30$
Accept H_0, since critical region is $\{0, 1, 2\}$.

(b) Let the mean number of vegetarian meals per 100 customers be λ.
H_0: $\lambda = 10$
H_1: $\lambda \neq 10$
Critical region is $\{0, 1, 2, 3\}$ and $\{18, 19, 20, \ldots\}$.

(c) Intended to be a 5% test, but actually is 2.5%.

INDEX